Ethical Practices for Sustainable Construction Digitalisation

Olugbenga Oladinrin • Douglas Aghimien
Steve Goodhew

Ethical Practices for Sustainable Construction Digitalisation

Springer

Olugbenga Oladinrin
Built Environment Department
University of Plymouth
Plymouth, UK

Steve Goodhew
Built Environment Department
University of Plymouth
Plymouth, UK

Douglas Aghimien
Department of Civil Engineering
Technology
University of Johannesburg
Johannesburg, South Africa

ISBN 978-3-032-00150-4 ISBN 978-3-032-00151-1 (eBook)
https://doi.org/10.1007/978-3-032-00151-1

This Springer imprint is published by the registered company Springer Nature Switzerland AG
The registered company address is: Gewerbestrasse 11, 6330 Cham, Switzerland

Preface

The fourth industrial revolution has significantly changed how industries operate and deliver products and services worldwide. This industrial revolution has brought with it ubiquitous emerging physical, biological, and digital technologies that are disrupting activities in sectors across the globe. The construction industry is not immune to the disruption of these technologies. As such, evidence of emerging physical and digital tools is being recorded within the industry. From solving the issues of poor cost, time and quality that have bedevilled the industry for a long while to ensuring clients are satisfied and workers are safe, digital technologies have proven effective in improving how the construction industry functions. Therefore, seeing the significant research attention diverse digital technologies are getting within the current academic discourse is unsurprising. For instance, building information modelling has garnered considerable attention in addressing several problems, such as the industry's salient design and collaboration issues. Also, the Internet of Things, big data analytics, cloud computing, drones, robotics, sensors, and a host of other technologies have been explored to improve the management and delivery of construction projects. Although significant attention has been garnered by these digital tools in construction, one aspect that has received less attention is ethics.

With the advent of diverse technologies designed to interact with humans in the delivery of projects, there is no denying that the issue of ethics has become important. Studies have noted that with continuous technological advancement, the line between a person and the physical world is continuously blurred, leading to ethical crises and significant impairments to the growth of institutional systems. Therefore, the role of ethics cannot be overlooked for the effective and sustainable digitalisation of the construction industry. This becomes important for the construction industry, which relies heavily on human input and human interaction with technology. Besides, with the dynamic and complex nature of construction, where projects generate a large amount of data, the ethical use of these data becomes essential. Regrettably, this concept of ethics has not gained much prominence in the quest for the digitalisation of construction industries globally. Furthermore, there is an absence of a roadmap for the ethical use of digital technologies in the digitalisation of the construction industry. Based on this knowledge, this book becomes timely as it provides an ethical practice framework for using digital tools to achieve effective and sustainable digitalisation in the construction industry.

Furthermore, the book is written to give value to readers as no existing study has focused on an ethical practice roadmap designed to ensure effective and sustainable digitalisation, particularly in the construction industry. It will benefit professionals such as architects, quantity surveyors, construction managers, project managers, engineers, urban and regional planners, and value managers to use emerging digital technologies in their daily activities. The book also strives to give a general perceptive of the constituents of ethical practices needed for developing and using technologies while delineating the key practices required by construction organisations, practitioners, and stakeholders to ensure they can effectively deploy digital technologies and sustain their attained digitalisation through adherence to ethical practices. It offers value to owners of construction organisations (contracting and consulting) who seek to attain better service delivery to their clients and gain a better competitive edge over their competitors through the ethical digitalisation of their business functions. The book also supports construction policymakers' and regulatory bodies' agenda to promote digitalisation and the ethical relationship between technologies and humans in the industry. The content can help emphasise the need for ethical use of digital technologies for construction project delivery and shape guidelines needed to drive this usage. Lastly, the book is designed to serve as a secondary tool for built environment researchers in academics and practice, seeking to expand the frontiers of knowledge on ethics in the digitalisation of the construction industry. To this end, the authors confirm that the text utilised in this work reflects original work and, where necessary, the material has benefited from relevant referencing.

Plymouth, UK — Olugbenga Oladinrin
Johannesburg, South Africa — Douglas Aghimien
Plymouth, UK — Steve Goodhew

Competing Interests The authors have no competing interests to declare that are relevant to the content of this manuscript.

Contents

About the Authors

Olugbenga Oladinrin is an Associate Professor (Teaching and Research) and Program Leader for the Built Environment at the School of Art, Design and Architecture, University of Plymouth, United Kingdom. He obtained his PhD in Construction Project Management from the Hong Kong Polytechnic University, Hong Kong. His research areas include construction ethics, construction innovation, sustainable construction, construction health and safety, new working practices, construction and demolition waste management, circular economy concept and quantity surveying practices.

Douglas Aghimien has a PhD in Engineering Management from the University of Johannesburg. He is a Fellow of the UK's Higher Education Academy and a Member of the Chartered Institute of Building. He is a researcher by the National Research Foundation of South Africa with interest in construction management, digitalisation, and sustainable construction. He has co-authored books on construction digitalisation, workforce management, maintenance engineering, and risk management. Dr Aghimien has continued to further the sustainable built environment discourse to optimise diverse construction processes and practices.

Steve Goodhew is a Professor of Environmental Building and REF coordinator of the School for Architecture, Design and Environment at the University of Plymouth, United Kingdom. He researches into the use of thermography to improve building performance, the use of ultra-low carbon natural materials and the thermal properties of earth building materials. Steve is a fellow of RICS and CIOB and is the author of Sustainable Construction Processes published by Blackwell.

Chapter 1
General Introduction

Introduction

Technology, like art, is a soaring exercise of the human imagination—Daniel Bell.

The advent of the fourth industrial revolution, characterised by cyber-physical systems integrating digital, physical, and biological technologies to improve production (Schwab, 2017), has brought rapid transformation to production in many industries (including construction). In construction, Kozlovska et al. (2021) noted that the influence of this industrial revolution can be seen in the form of digitalisation and industrialisation of the construction industry and its processes. As such, digitalisation has become synonymous with the fourth industrial revolution, in which organisations use digital technologies to offer services more smartly (Aghimien et al., 2021). Gartner (2016) noted that organisations achieve digitalisation through digital technologies deployed to enhance their business models and provide new revenue and opportunities that add value to their businesses. Additionally, Bahl (2015) described digitalisation as an innovation that links technology, business strategy, devices, data science, and design to change business processes. This digitalisation allows organisations to deliver their services more easily and faster through cutting-edge digital technologies that afford them a competitive advantage.

This concept of digitalisation in construction has continued to gain prominence due to its inherent benefits. As such, studies have continued to emerge on the deployment of digital technologies like building information modelling (BIM), artificial intelligence (AI), augmented reality (AR), virtual reality (VR), digital twin, big data analytics (BDA), blockchain, cloud computing and drones, among others within the construction industry (Chen et al., 2022; Elghaish et al., 2021; Eze et al., 2024; Kozlovska et al., 2021; Won et al., 2022). Others have explored the challenges and benefits of adopting these digital technologies (Delgado et al., 2019; Oke et al., 2021) and the trends in digital technology advancements (Pan & Pan, 2020). However, the discourse around the ethical use of technologies in construction has received less attention. This is surprising as one of the major issues hindering

O. Oladinrin et al., *Ethical Practices for Sustainable Construction Digitalisation*, https://doi.org/10.1007/978-3-032-00151-1_1

project success in the construction industry is poor ethical conduct of the industry's professionals and stakeholders.

In simple terms, ethics is the moral principles that govern an individual or group of people's behaviour or how they conduct themselves in certain situations. It has been described as right and wrong acts and the consequent reactions. Also, ethics has been viewed as the study of morals and the norms that society acknowledges as wrong or right and good or bad (Crane & Matten, 2010; Velasquez et al., 2010). Cartlidge (2011) described it as the science of moral principles or codes. Velasquez et al. (2010) further noted that ethics is rooted in standards of right or wrong that guide people on what they ought to do, especially in rights, obligations, fairness or virtues. In the construction industry, ethics has been described as standards that influence workers' and stakeholders' behaviours (Oladinrin & Ho, 2016a). It has been observed that there exists a complex interplay between diverse stakeholders within the construction industry, and if not properly managed or guided by set codes of conduct, it can lead to conflict of interests and ethical issues (Loosemore & Lim, 2017). This has necessitated diverse ethical considerations, policies, and codes of conduct designed to ensure professionals and stakeholders can navigate ethical issues in the industry (Olatunji et al., 2016). In fact, Oladinrin and Ho (2016b) noted the need for ethical behaviour to become a priority within every construction organisation and to achieve this, there must be processes to align employee behaviour with the organisation's ethical standards.

Regrettably, the construction industry has been criticised for its poor ethical standards. Amoah and Steyn (2023) noted that there are many unethical and corrupt practices that deter effective management and execution of construction projects in many construction industries globally. Wold et al. (2019) have described these unethical practices as questionable behaviours. These behaviours contradict professional norms and are usually frowned upon within the industry. Studies have continued to showcase the unethical behaviours of construction professionals and stakeholders, and this has been noted as a crucial problem in the construction industry (Oladinrin et al., 2023; Shah & Alotaibi, 2017). Haupt and Harinarain (2016) noted that there have been instances of cartels being formed by organisations to take sole control of a sector or geographical region. This allows these cartels to monopolise the market and inflate tender prices as they see fit. The result is an unnecessarily high cost of construction and clients not getting value for their money as they pay exorbitantly for their projects. This problem has been a global issue with corruption, bribery, bid shopping, nepotism, cover pricing, bid cutting, and conflict of interest, among others, have been noted as some of the drawbacks of effective construction project delivery in the global construction industry (Adnan et al., 2012; Ameyaw et al., 2017; Amoah & Steyn, 2023; Ogunmankinde et al., 2019; Olatunji et al., 2016). Studies have noted that unethical practices harm the construction industry, economy and human resources by creating a bad reputation, which is an unattractive trait for investors. Moreover, these practices adversely affect construction performance, leading to increased project costs and reduced quality. They also inhibit economic development while creating social inequality (Aigbavboa et al., 2016; Ameyaw et al., 2017; Maseko, 2017; Oladinrin & Ho, 2016a; Olatunji et al., 2016).

The importance of having clear ethical codes and imbibing ethical practices in the construction industry transcends the delivery of construction projects and the use of emerging digital technologies for construction digitalisation. Ethics in the development and use of digital technologies has been described as the process of applying reasoning to practical issues involving the application of technology (Green, 2024). It is the moral principles and standards guiding individuals' and organisations' ethical conduct and responsibilities within the digital landscape (Smowltech, 2024). Braunack-Mayer et al. (2012) described it as simply the study of morals and the ethical consequences of developing and implementing emerging technologies. In essence, ethics in technology development and use entails prioritising ethical considerations for the fair and just use of technologies. According to Green (2024), this concept of ethics in the context of technology usage is gaining prominence because emerging technologies provide humans with greater power and choices than ever before. Also, technological advancement has made it challenging to adequately predict the effect of any technology; as such, it is important to have good ethical guidelines for their usage.

In the quest for construction digitalisation, there is the deployment of emerging digital technologies such as BIM, IoT, BDA, cloud computing, sensors, AR and VR, among others, that allow the capture and management of massive volumes of data (Aghimien et al., 2021). This begs the question of how these data are stored, shared, and used. Moreover, there are ethical concerns about data ownership and the potential for misuse or breaches, and this has necessitated the development of frameworks and policies to protect stakeholder interests (Oladinrin et al., 2023). Ethical considerations are necessary for construction organisations, professionals, and stakeholders to be able to effectively deploy these emerging technologies ethically and sustainably. Moreover, the current rise in the use of digital technologies in the global construction industry needs to be sustained (Aghimien et al., 2021). If ethical considerations are not implemented, construction organisations will face the issue of privacy invasion and violation, which might lead to dire consequences. Exploitation, risk of data breaches, poor accountability and responsibility, privacy and security issues, lack of transparency, regulatory compliance problems, accuracy problems, and ethical washing, among others, is bound to impede the digitalisation journey of these organisations if not carefully addressed (Sacks et al., 2020; Schoentgen & Wilkinson, 2021; Xue et al., 2022; Zhao, 2024).

It has been noted that through ethical considerations in using digital technologies, organisations can build trust, which can be a basis for competitiveness in attracting and retaining construction clients (Joseph, 2023; Veinhardt, 2022). Construction organisations can also avoid legal and financial risks associated with non-compliance with existing legislations and regulations or wrongful use of people's data (Jemmy, 2024). Moreover, ethical practices in using digital technologies offer sustainability of organisational business. It has been observed that organisations can attain long-term sustainability as ethical considerations align with sustainable practices that consider efficient use of resources and elimination of waste in construction. Also, by prioritising the social impacts of digital technologies, construction organisations can contribute to the well-being of communities and the environment (Martínez-Peláez et al., 2023).

Albeit the importance of ethical practices in the use of technologies and the ethical challenges identified, a review of extant literature revealed a shortage of studies on ethical practices in digital technology adoption for effective and sustainable digitalisation of construction organisations. Addressing this knowledge gap becomes timely as proposing a pathway for the ethical use of digital technologies promises immense benefits for these organisations. Based on this notion, this book was designed to provide a conceptualised ethical practice framework that will engender sustainable digitalisation within construction organisations.

Objectives of the Book

The importance of ethical practices in deploying digital technologies for the digitalisation of construction organisations cannot be overemphasised. The consequences of failing to acknowledge ethical practices in the use of these technologies can be severe for construction organisations, with issues including loss of reputation, which can lead to loss of future clients and jobs, infringement on existing legislations and regulations with its attributed cost implications, reduced productivity and high risk of delivery project poorly all being potential problems. The absence of a roadmap for the ethical use of digital technologies by construction organisations, professionals and stakeholders seeking to attain effective and sustainable construction digitalisation birthed this book. To this end, the objective of this book was to provide a conceptualised ethical practice framework designed to offer effective and sustainable digitalisation of construction organisations. To achieve this objective, the book X-rayed the anatomy of the construction industry vis-à-vis the industry's readiness to be digitalised. Furthermore, the book explored the concept of sustainable construction in relation to digitalisation as well as the ethics and digitalisation nexus. In developing the proposed framework, the book draws from existing digital ethics and ethics of technology frameworks. Ultimately, the book proposes four construct digitalisation and six ethical consideration frameworks designed to offer sustainable construction digitalisation, better competitiveness, increased innovativeness, and the opportunity to become a digital disruptor, among other benefits.

Contribution and Value

The value of the book can be seen from several folds. Firstly, the book is timely as it provides a collection of knowledge for academics, organisations, and regulatory bodies to use in their quest for sustained construction digitalisation, which is achieved through proper ethical practices. The book fills the knowledge gap regarding the existence of an ethical practice framework specifically for the sustainable digitalisation of construction organisations. The provided conceptual framework gives a direction for attaining digitalisation by providing four core digitalisation dimensions that construction organisations seeking to be digitalised must consider. These dimensions are technology, people, process and strategy, forming the foundation for organisations to build their digitalisation journey. The conceptual framework further proposes six ethical considerations that must guide construction organisations' quest for sustained digitalisation. Considering accountability, equity and fairness, privacy, safety and security, sustainability, and transparency, construction organisations will be able to promote ethical use of digital technologies and, in the process, gain sustainable digitalisation along with other inherent benefits. Moreover, the proposed framework can be useful to policymakers within the construction industry. Using the framework, industry regulatory and professional bodies and other relevant policymakers can be aware of the ethical considerations with which their members and industry practitioners must abide. It will further help them align their existing regulations to incorporate these ethical considerations and help them ensure that organisations are ethically employing the use of technologies in the delivery of the industry's products. The submission of the book also serves as an excellent platform for future built environment researchers in academics and practice seeking to explore the ethical use of emerging digital technologies. It gives a basis for further research in this area, particularly in countries where such study is non-existent.

Structure of the Book

The book is divided into eight chapters, with Chap. 1 presenting a general introduction to the book. Chapter 2 delineates the concepts of digitalisation, ethics, and sustainability with a view to providing preliminary insights into these concepts. This chapter gives insight into the fourth industrial revolution and clarifies the differences between digitisation, digitalisation and digital technologies. The chapter further gives an overview of ethics and ethics of technology, sustainability, and the triple bottom line. Chapter 3 explored digitalisation in the construction industry by illuminating the nature of the global construction industry and the different digital technologies in construction. Chapter 4 was designed to unearth sustainable construction practices in the digital era. The chapter delineates the concept of sustainable construction and its practices. The chapter further revealed the different technologies that can promote sustainable construction, and the different areas of focus of past research in sustainable construction were unearthed through bibliometric review. Chapter 5 explored ethical practice in digitalisation. It explored the concept of ethics, ethics in construction, and ethical practices and digitalisation. The chapter further explored the impact of ethics on digitalisation, the factors influencing ethics in digitalisation and ethical concerns about emerging technologies. Chapter 6 presents ethical practices theories, frameworks and models for sustainable construction digitalisation. This chapter presents 11 relevant digital ethics and ethics of technology frameworks reviewed to unearth the key ethical considerations for the proposed ethical practice framework. Chapter 7 conceptualised the effective, ethical practices for sustainable construction digitalisation. The chapter presents the digitalisation dimensions and ethical considerations needed for effective and sustainable digitalisation of construction organisations. Chapter 8 concludes the book by highlighting the salient points in the previous chapters and showcasing the practical and theoretical contributions of the book as well as its recommendations.

Summary

This first chapter provides the readers with the idea behind the conception of this book. The rise in the use of technologies in the delivery of construction projects has necessitated the need to put measures in place to ensure that these technologies are used ethically. The storing and analysing of individual and project data, technology-human interaction, and use of technologies for crucial decision-making all call for ethical practices to be adhered to. These concerns, coupled with the absence of an ethical practice framework for effective and sustainable digitalisation tailored for the construction industry, birthed this book. This first chapter also details the objectives of the book, its value, and its structure. The next chapter describes the concept of digitalisation, ethics and sustainability.

References

Adnan, H., Hashim, N., Yusuwan, N. M., & Ahmad, N. (2012). Ethical issues in the construction industry: Contractor's perspective. *Procedia - Social and Behavioural Sciences, 35*, 719–727.

Aghimien, D. O., Aigbavboa, C. O., Oke, A. E., & Thwala, W. D. (2021). *Construction digitalisation—A capability maturity model for construction organisations*. Routledge.

Aigbavboa, C., Oke, A., & Tyali, S. (2016). Unethical practices in the South African construction industry. In F. Emuze (Ed.), *Proceedings of the 5th construction management conference* (pp. 15–22). Nelson Mandela Metropolitan University.

Ameyaw, E. E., Pärn, E., Chan, A. P. C., Owusu-Manu, D., Edward, D. J., & Darko, A. (2017). Corrupt practices in the construction industry: Survey of Ghanaian experience. *Journal of Management in Engineering, 33*(6), 1–11.

Amoah, C., & Steyn, D. (2023). Barriers to unethical and corrupt practices avoidance in the construction industry. *International Journal of Building Pathology and Adaptation, 41*(6), 85–101.

Bahl, M. (2015). *Asia rising: Digital driving*. Cognizant Centre for the Future of Work. Retrieved February 2, 2024, from http://www.futureofwork.com

Braunack-Mayer, A., Street, J. M., & Palmer, N. (2012). Technology, ethics of: Overview. In *Encyclopedia of applied ethics* (pp. 321–327). Academic Press.

Cartlidge, D. (2011). *New aspects of quantity surveying practice*. Spon Press.

Chen, X., Chang-Richards, A. Y., Pelosi, A., Jia, Y., Shen, X., Siddiqui, M. K., & Yang, N. (2022). Implementation of technologies in the construction industry: A systematic review. *Engineering, Construction and Architectural Management, 29*(8), 3181–3209.

Crane, A., & Matten, D. (2010). *Business ethics: Managing citizenship and sustainability in the age of globalisation*. Oxford University Press.

Delgado, J. M. D., Oyedele, L., Ajayi, A., Akanbi, L., Akinade, O., Bilal, M., & Owolabi, H. (2019). Robotics and automated systems in construction: Understanding industry-specific challenges for adoption. *Journal of Building Engineering, 26*, 1–11.

Elghaish, F., Matarneh, S., Talebi, S., Kagioglou, M., Hosseini, M. R., & Abrishami, S. (2021). Toward digitalisation in the construction industry with immersive and drone technologies: A critical literature review. *Smart and Sustainable Built Environment, 10*(3), 345–363.

Eze, E. C., Aghimien, D. O., Aigbavboa, C. O., & Sofolahan, O. (2024). Building information modelling adoption for construction waste reduction in the construction industry of a developing country. *Engineering, Construction and Architectural Management, 31*(6), 2205–2223.

Gartner. (2016). *Digitalisation*. Retrieved December 11, 2023, from https://www.gartner.com/en/information-technology/glossary/digitalisation

Green, B.P. (2024). *What is technology ethics?* Retrieved February 2, 2024, from https://www.scu.edu/ethics/focus-areas/technology-ethics/#:~:text=Technology%20ethics%20is%20the%20application,t%20have%20to%20make%20before

Haupt, T. C., & Harinarain, N. (2016). The image of the construction industry and its employment attractiveness. *Acta Structilia, 23*(2), 79–108.

Jemmy, S. (2024). Integration of business ethics in managerial decisions implications for companies' long-term growth. *International Journal of Social and Education, 1*(6), 1689–1702.

Joseph, A. (2023). The importance of business ethics: Building trust and sustainability. *Business Studies Journal, 15*(S2), 1–2.

Kozlovska, M., Klosova, D., & Strukova, Z. (2021). Impact of industry 4.0 platform on the formation of construction 4.0 concept: A literature review. *Sustainability, 13*, 2683.

Loosemore, M., & Lim, B. T. H. (2017). Linking corporate social responsibility and organisational performance in the construction industry. *Construction Management and Economics, 35*(3), 90–105.

Martínez-Peláez, R., Ochoa-Brust, A., Rivera, S., Félix, V. G., Ostos, R., Brito, H., Félix, R. A., & Mena, L. J. (2023). Role of digital transformation for achieving sustainability: Mediated role of stakeholders, key capabilities, and technology. *Sustainability, 15*(14), 11221.

Maseko, C. (2017). Literature on theory and practice on unethical practices in the construction of projects: A case of an emerging economy. *Risk Governance and Control: Financial Markets and Institutions, 7*(4), 214–224.

Ogunmankinde, O.E., Sher, W. & Maund, K. (2019). Challenges of the Nigerian construction industry: A systematic review. In: *IISBE forum of young researchers in sustainable building*, Czech Republic, Prague, 1 July.

Oke, A. E., Farouk Kineber, A., Abdel-Tawab, M., Abubakar, A. S., Albukhari, I., & Kingsley, C. (2021). Barriers to the implementation of cloud computing for sustainable construction in a developing economy. *International Journal of Building Pathology and Adaptation, 41*(5), 988–1013.

Oladinrin, O. T., Arif, M., Rana, M. Q., & Gyoh, L. (2023). Interrelations between construction ethics and innovation: A bibliometric analysis using VOSviewer. *Construction Innovation, 23*(3), 505–523.

Oladinrin, O. T., & Ho, C. M. F. (2016a). Embeddedness of codes of ethics in construction organisations. *Engineering, Construction and Architectural Management, 23*(1), 75–91.

Oladinrin, O. T., & Ho, C. M. F. (2016b). Enabling ethical code embeddedness in construction organisations: A review of process assessment approach. *Science and Engineering Ethics, 22*, 1193–1215.

Olatunji, S. O., Oke, A. E., Aghimien, D. O., & Ogunwoye, O. S. (2016). Implementation of code of ethics among quantity surveying firms in Nigeria. *Developing Country Studies, 6*(5), 71–76.

Pan, M., & Pan, W. (2020). Stakeholder perceptions of the future application of construction robots for buildings in a dialectical system framework. *Journal of Management in Engineering, 36*(6), 846.

Sacks, R., Girolami, M., & Brilakis, I. (2020). Building information modelling, artificial intelligence and construction tech. *Developments in the Built Environment, 4*, 100011.

Schoentgen, A., & Wilkinson, L. (2021). Ethical issues in digital technologies. In *23rd biennial conference of the international telecommunications society (ITS): "Digital societies and industrial transformations: Policies, markets, and technologies in a post-Covid world"*. International Telecommunications Society (ITS).

Schwab, K. (2017). *The fourth industrial revolution* (1st ed.). Crown Business.

Shah, R. K., & Alotaibi, M. (2017). A study of unethical practices in the construction industry and potential preventive measures. *Journal of Advanced College of Engineering and Management, 3*, 55–77.

Smowltech. (2024). Digital ethics: Definition and examples. Retrieved August 12, 2024, from https://smowl.net/en/blog/digital-ethics/#:~:text=Digital%20ethics%20refers%20to%20the,and%20society%20as%20a%20whole

Veinhardt, J. (2022). The dilemma of postmodern business ethics: Employee reification in a perspective of preserving human dignity. *Frontiers in Psychology, 373*, 813255.

Velasquez, M., Andre, C., Shanks, T. & Meyer, M.J. (2010). *What is ethics?* Retrieved August 12, 2024, from https://www.scu.edu/ethics/ethics-resources/ethical-decision-making/what-is-ethics/

Wold, J., Lædre, O., & Lohne, J. (2019). Questionable practice in the processing of building permits in Norway. In C. Pasquire & F. R. Hamzeh (Eds.), *Proceedings of the 27th annual conference of the International Group for Lean Construction (IGLC)* (pp. 1151–1162). International Group for Lean Construction.

Won, D., Hwang, B., & Samion, N. K. (2022). Cloud computing adoption in the construction industry of Singapore: Drivers, challenges, and strategies. *Journal of Management in Engineering, 38*(2), 1–14. https://doi.org/10.1061/(ASCE)ME.1943-5479.0001001

Xue, X., Tan, X., Huang, Q., Zhu, H., & Chen, J. (2022). Exploring the innovation path of the digital construction industry using mixed methods. *Buildings, 12*(11), 1840.

Zhao, L. (2024). Construction information security management system based on data sharing algorithm. *Intelligent Decision Technologies, 18*, 2775–2788.

Chapter 2
Digitalisation, Ethics, and Sustainability

Introduction

The construction industry globally is impacted by the fourth industrial revolution, which is driven by emerging technologies. This industrial revolution is characterised by cyber-physical systems, and it seeks to merge the digital, physical, and biological spheres into a single sphere for effective and efficient production (Lu, 2017). An emergence of this industrial revolution is digitalisation, which entails converting manual processes and analogue information into digital processes and formats through digital technologies to change how businesses operate (Gartner, 2016; Monton, 2022). This process of digitalisation offers better storage of information, efficient collaboration and decision-making, increased efficiency in the organisation's processes, and the opportunity to offer value for money (Aghimien et al., 2021). This makes exploring the concept of digitalisation crucial for the chapter. However, the use of these digital technologies to attain digitalisation brings ethical concerns that might impact the sustainability of construction services. As such, understanding the concept of ethics, especially regarding technology development and use, is also important. Furthermore, sustainability, which entails being able to meet the needs of the current and future generations, is also an important concept worth exploring in the quest to determine ethical practices for sustainable digitalisation in construction organisations. This is because the use of digital technologies has a societal, economic, and environmental impact, which, if not addressed, might affect organisations' ability to sustain their digitalisation process. To this end, this chapter delineates the concepts of digitalisation, ethics, and sustainability as they relate to the objective of this book.

O. Oladinrin et al., *Ethical Practices for Sustainable Construction Digitalisation*, https://doi.org/10.1007/978-3-032-00151-1_2

The Road to the Fourth Industrial Revolution and Digitalisation

The fourth industrial revolution, popularly called Industry 4.0, began with the first industrial revolution, which occurred in the late 1700s and mid-1800s. During this first industrial revolution, a mechanised production economy driven by water and steam engines for manufacturing was birthed. In the early 1990s, there was a shift from the first to the second industrial revolution, dubbed the 'technological revolution' due to the series of scientific discoveries, mass production, and industrialisation that was driven by electrical energy. In the late 1960s, the third industrial revolution, also known as the 'information age', started. This industrial revolution saw the advent of information technologies and electronics for industrial production, leading to the development of automated and optimised manufacturing methods. Semiconductors, mainframe computers, personal computers, and the internet were some of the developments from this revolution, earning it the moniker of the 'computer age' or the 'digital revolution'. The information age serves as the bedrock for the fourth industrial revolution. This latest revolution is characterised by the integration of emerging technologies, such as the Internet of Things (IoT), artificial intelligence (AI), big data analytics (BDA), robotics, and additive manufacturing, to drive greater efficiency, innovation, and transformation across various industries (Lu, 2017; Schwab, 2017).

The term 'Industry 4.0' was first established at the Hannover Fair in 2011 in Germany. Since its pronouncement, the Industrial Revolution has gained traction within academic and industry discourse globally (Sung, 2018). According to Bahrin et al. (2016), Industry 4.0 is an emerging structure for manufacturing and logistics systems to gain easy access to information and communication networks to align automated business processes and production. Rüßmann et al. (2015) earlier noted that by embracing Industry 4.0, higher efficiency and changes in traditional production systems can be achieved. In the process, the relationships among people in the manufacturing system and between man and machines can improve. However, the scope of Industry 4.0 extends far beyond the basic development of smart and connected systems. It is characterised by groundbreaking advancements across various emerging technologies, including gene sequencing, nanotechnology, renewable and reusable materials, and quantum computing. The ability of these technologies to seamlessly interact with one another, as well as across the physical, digital, and biological domains through cyber-physical systems (CPS), is what truly sets Industry 4.0 apart as a significant advancement from the third industrial revolution (Schwab, 2017).

Rather than being limited to improving manufacturing and logistics processes, Industry 4.0 is ushering in transformative changes that can reshape entire industries and fundamentally alter how we live, work, and interact with the world around us. The convergence and synergistic integration of these cutting-edge technologies drive unprecedented innovation, efficiency, and disruption across a wide spectrum of economic and societal spheres. As such, Industry 4.0 represents a profound transformation in how humans exist and interact. It is a new phase in human development, made possible by remarkable technological advancements paralleling the first three industrial revolutions. These innovations fuse the physical, digital, and biological domains, creating immense promise and potential risks. The extraordinary speed, breadth, and depth of this revolution are compelling humans to rethink how countries develop, how organisations create value, and even our conception of what it means to be human. Unlike previous industrial eras, Industry 4.0 is evolving exponentially and disrupting nearly every industry globally, driving massive, non-linear change at unprecedented velocities (Aghimien et al., 2021; Schwab, 2017).

One of the core components of Industry 4.0 is IoT—the extensive network of connected physical devices worldwide that collect and share data (Ashton, 2009). IoT allows unprecedented integration between physical and virtual realms, giving rise to smart systems that can adapt independently to environmental changes (Lasi et al., 2014). Similarly, Industry 4.0 is driven by AI with systems that can execute tasks requiring human intelligence, including visual perception, speech recognition, decision-making, and language translation. The importance of AI has led to its deployment in diverse sectors, including healthcare, finance, manufacturing, and construction. As the use of AI becomes popular, its sophistication continues to increase, giving it the ability to carry out tasks that previously demanded human cognition (Russell & Norvig, 2016). With AI comes BDA, which offers the ability to analyse massive datasets and discover patterns, correlations, and trends. Mayer-Schönberger and Cukier (2013) described it as a critical requirement for many emerging technologies. Advancements in robotics are automating manual tasks and enhancing human capabilities. Brynjolfsson and McAfee (2014) submitted that collaborative robots work alongside humans, bringing new levels of efficiency and safety to a variety of tasks. Also, additive manufacturing, also known as three-dimensional (3D) printing, is transforming production processes, allowing for the creation of complex structures with reduced material waste, with applications spanning aerospace to biotechnology (Gibson et al., 2015). The convergence of technologies is a hallmark of Industry 4.0, enabling new interactions between digital and physical systems, such as digital twins—virtual replicas of the physical world used for simulation and control (Grieves & Vickers, 2017).

Industry 4.0 undoubtedly is having a significant impact on the job market. While some jobs may be lost due to automation and the shift in the division of labour between humans and machines, others will be created, and many existing roles will undergo dramatic changes (Aghimien et al., 2023; Frey & Osborne, 2017). Education systems are being challenged to adapt to the needs of this revolution, with a growing emphasis on science, technology, engineering, and mathematics (STEM) education and the development of skills such as creativity, critical thinking, and collaboration (Schwab, 2017). However, the transformative nature of Industry 4.0 also raises ethical and governance concerns. The rapid advancement of new technologies may outpace the existing legal and regulatory frameworks, requiring policymakers and stakeholders to re-examine crucial issues such as privacy, intellectual property, and liability in the context of these emerging innovations (Allen & Chan, 2017). Addressing these ethical and regulatory challenges will be crucial to ensure the responsible and equitable implementation of Industry 4.0 technologies, while also harnessing their full potential to drive economic and societal progress. Therefore, Industry 4.0, with its convergence of cutting-edge technologies, presents both opportunities and challenges. As the workforce and education systems adapt to these changes, equal attention must be given to establishing appropriate ethical guidelines and governance structures to safeguard the interests of individuals, businesses, and society.

Digitisation, Digitalisation, and Digital Technologies

Industry 4.0 is underpinned by a convergence of emerging technologies across the physical, digital, and biological domains. These technological advancements are the driving force behind this industrial revolution, as without them, its inherent benefits cannot be fully realised (Schwab, 2017). While continuous progress is being made in each of these three technology domains, the breakthroughs and the ability of these technologies to interconnect are what distinguish the Industry 4.0 from previous industrial revolutions. Majorly Industry 4.0 is rooted in the transformative power of software and digital technologies (Aghimien et al., 2021). In the manufacturing sector, where the concept of the Industry 4.0 originated, the integration of technologies such as IoT, Industrial IoT, cloud-based systems, and smart manufacturing have emerged as prominent drivers for successful production. These technologies have led to the creation of smart factories that are revolutionising the manufacturing industry globally (Bédard-Maltais, 2017; Erol et al., 2016). According to Erol et al. (2016), the internet and CPS serve as a foundation for the seamless integration of human and machine agents, materials, products, production lines, and processes, leading to more intelligent, connected, and agile value chains. In essence, the concept of Industry 4.0 inherently relies on digitalisation, as it is the driving force that enables the realisation of this industrial revolution (Bienhaus & Haddud, 2018). However, it is important to note that while the terms 'digitisation', 'digital technologies', and 'digitalisation' are sometimes used interchangeably in the literature, they have some distinctness (Aghimien et al., 2021).

Digitisation

Digitisation is the process of converting analogue information into a digital format, represented by a series of zeros and ones. This process allows information to be expressed in various ways, on different media, and through diverse systems. All forms of data can be digitised, including text, graphics, images, and audio (Brennen & Kreiss, 2014; Verhulst, 2002). In essence, digitisation in an organisation is the conversion of analogue-based elements, such as customer files, repair manuals, ledgers, and other materials, into a more readily accessible and transportable digital format for easy access and sharing through digital devices. This transformation from analogue to digital enables organisations to streamline their operations, improve data management, and leverage the advantages of digital technologies (Aghimien et al., 2021; Irniger, 2017). The process of digitisation is a fundamental to the broader concept of digitalisation involving the integration of digital technologies into all aspects of an organisation's activities, leading to fundamental changes in how the organisation operates and delivers value to its stakeholders. Digitisation is the mainstay for digitalisation, allowing organisations to harness the full potential of digital tools and capabilities to drive innovation, enhance efficiency, and adapt to the evolving digital landscape.

Digitalisation

Digitalisation is more concerned with how the conversion of analogue information into digital formats (digitisation) shapes people's social and economic lives rather than the technical process of converting zeros and ones. Crampton (2016) described digitalisation as using digital technologies to transform existing business models and generate new streams of income while producing value-generating opportunities. In essence, digitalisation involves adopting and integrating digital technologies within an organisation's operations (Kalavendi, 2017). Digitalisation is widely regarded as one of the most transformative technological advancements, with impacts that will be felt across all levels of society (Leviakangas et al., 2017). It is believed that digitalisation has permeated every aspect of human life, from simple personal devices to complex industrial systems (Dall'Omo, 2017). It incorporates digital technologies into everyday life by digitising anything that can be converted into a digital format (Ochs and Riemann (2018). Leviakangas et al. (2017) further highlighted that digitalisation enables the creation of unique technological foundations for metasystems of products, leveraging intelligent sensors, robotics, and automation. Thus, while digitisation is the transformation of objects and processes that were previously analogue, digitalisation is the deployment of these transformed processes into the organisation's business operations using digital technologies.

Digital Technologies

Digital technologies are the foundation of Industry 4.0 since everything in this industrial era is driven by digital capabilities (Schwab, 2017). Johnstone et al. (2024) noted that digital technologies are electronic tools, systems, devices, or resources that can create, store, and process data. They rely on microprocessors and operate through binary computational codes. Digital technologies encompass a wide range of computerised devices and applications that are dependent on electronic devices, including mobile phones, video cameras, tablets, laptops, desktop computers, the internet, and communication satellites (Anjos-Santos et al., 2016; Pullen, 2009). Dimick (2014) further categorised digital technologies into software, information technology (IT), and communications. They can be used to collect, store, analyse, and transmit vast amounts of data, facilitating the development of innovative applications, services, and business models. Integrating these digital capabilities across industries and sectors is a hallmark of the Industry 4.0, driving increased automation, connectivity, and intelligence within various systems and processes. The ubiquitousness and transformative potential of digital technologies are central to the ongoing evolution of the industrial landscape. This is because organisations continue to seek to harness these tools to enhance efficiency, increase productivity, and unlock new opportunities for growth and value creation.

With the advent of diverse digital technologies designed to interact with humans in the delivery of projects, there is no denying that the issue of ethics has become important. Studies have noted that with continuous technological advancement, the line between a person and the digital world is continuously blurred, leading to ethical crises and significant impairments to the growth of institutional systems. Therefore, the role of ethics cannot be overlooked for the successful digitalisation of the construction industry.

Ethics and Ethics of Technology: An Overview

From a generic perspective, ethics is the science of moral principles or codes. According to Robinson et al. (2007), ethics is a philosophical view of right or wrong in human activities and what principles should govern these activities. Ethics can also be viewed from the perspective of moral beliefs and practices (Mansfield, 2008). Longstaff (1999) noted that a professional's ethical environment comprises choices, decisions, and practical application of shared core values and principles. As such, ethics can provide a moral compass to navigate through difficult professional environments (Cartlidge, 2011). Society of Chartered Surveyors (2006) described ethics as a set of principles that extends beyond the formal code of conduct that guides professionals in resolving issues that might impact their professional interest, client, or the community. Therefore, good ethical behaviour is germane to professionalism. While diverse professions within the built environment have specific code of conduct that governs their members, the general expectations cut across integrity, honesty, trustworthiness, respect, fairness, equality, objectivity, competence, care, confidentiality, and acting in the public's interest (Chartered Institute of Building, 2018; Olatunji et al., 2016; Royal Institute of Chartered Surveyors, 2021). The importance of instilling good ethical behaviour and standards within built environment professionals. This is because the built environment, which comprises architecture, engineering, and construction industry, has been known for its common unethical practices (Ameyaw et al., 2017; Paul et al., 2021).

The construction sector globally has been plagued by unethical behaviour, which has significantly undermined its integrity, as posited by Bowen et al. (2007). Across most countries, the building industry is viewed negatively in terms of ethics, with concerns ranging from corruption, malpractice, and mismanagement to exploitative business tactics and a lack of accountability for the well-being of its employees and other stakeholders (Ameyaw et al., 2017; Bowen et al., 2007; Oladinrin et al., 2023). Recent events across diverse countries have brought to light issues of shoddy construction work, the collapse of built structures in various nations, instances of modern slavery, and inadequate worker protection within the construction industry and this has raised concerns in the mind of the public and has affected the public's perception of the industry (Oladinrin et al., 2023). Estimates suggest that up to one-third of government investments in construction are wasted due to corruption and incompetence (Egbu & Ofori, 2018). This poor ethical culture of the construction industry requires significant attention, particularly as good ethical standards can help avoid corrupt activities and promote stakeholder confidence (Oladinrin & Ho, 2014).

In the quest to promote ethical use of emerging digital technologies, Liu and Wang (2020) warned about the mix-up between professional ethics and ethics of technology. It was noted that while both concepts are related, they are not the same and must be treated as individual entities in the quest for ethical use of technologies. While professional ethics deals with the science of moral principles or codes (Cartlidge, 2011), ethics of technology deals with applying ethical reasoning to the practical issues surrounding technology (Green, 2024). According to Smowltech (2024), ethics of technology encompasses the moral principles and standards that guide the ethical conduct and responsibilities of individuals, organisations, and governments in the digital landscape. It includes important factors such as fairness, transparency, privacy, security, and the influence of technology on individuals and society. Liu and Wang (2020) described the ethics of technology as the man-nature nexus, the holistic natural and social properties of technology, as well as the exploitation of man's intrinsic power using technology. Green (2024) noted that ethics of technology is becoming topical among researchers and professionals today because emerging technologies are providing humans with greater power and choices than ever before. Unlike in the past, when human actions were limited by their capabilities, today, man must navigate the responsibility of making decisions guided by ethical judgement and self-restraint. Smowltech (2024) submitted that ethics of technology strives to address issues of data privacy and misuse, algorithmic bias, cybersecurity, digital divide, and online harassment, among others. With the constant advancement in technologies, it is difficult to precisely predict a technology's effect. As such, having a good mindset in using these technologies is important.

Ashok et al. (2022), in a comprehensive review, submitted that ethics in the emerging digital world cut across several domains. This includes 'technoethics', which is concerned with how emerging technologies are developed and how they impact the ability of humans; 'roboethics', which is concerned with understanding and regulating the ethical implications and consequences of robotics while ensuring moral designs, development and applications of robots; information ethics also known as digital ethics which explores the correlation between creation, organisation, dissemination, and use of information, as well as the ethical standards and moral codes guiding human conduct in society; AI ethics which explores the responsibility, transparency, auditability, incorruptibility, predictability; the morality of machines.

Evidently, the concept of ethics transcends the morals and behaviours of professionals in the conduct of their day-to-day activities. With constant technological and innovative advancement, the ethical use of emerging digital technologies and innovations becomes crucial. Anton et al. (2021) noted a neglect of the concept of ethics in the use of emerging digital technologies like AI, particularly for surveillance. The same is the case in the construction industry, where despite the myriads of researches conducted on the application of different emerging technologies to solve crucial problems within the industry, the dimension of ethics has received little attention. This is despite the growing concern over ethical issues in adopting emerging technologies and innovations within the industry (Khodabakhshian, 2024; Oladinrin et al., 2023).

Consequently, the need for a robust discourse on the ethical use of digital technologies in the construction industry cannot be overemphasised. Khodabakhshian (2024) suggested that as technology continues to advance with an emphasis on human-centric concepts, several ethical elements become paramount. These are developing robust ethical standards and regulations that are necessary to shape the development and application of emerging technologies, creating ethical systems that go beyond just technological functionality but promote fair and moral decisions, and assessing the productivity and success of projects based on their social impacts, and not just financial profit. Simply put, as technology advances, it is important to prioritise the ethical frameworks, systems, and evaluation criteria that prioritise human well-being and social impact.

McAleenan et al. (2019) noted that although technological advancements in the form of smart wearables and implantable devices are effective for monitoring workers' location, health and well-being on construction sites, the ethical usage of these technologies is not given adequate attention. While emerging technologies are being utilised to enable efficient delivery of construction projects and improve workers' productivity, the ethical considerations underpinning the research and development of these technologies, as well as the morality in their practical application within the industry, require adequate attention. Also, Lin, and Xu (2021) explored data ethics based on big data technology and noted that, despite the benefits inherent in the use of this technology, ethical problems such as information security, privacy disclosure, and data gaps are apparent. Therefore, having data ethics in place becomes crucial. This is because, from the perspective of the individuals whose data is being collected and used, their awareness and understanding of their rights are limited. More so, from the perspective of organisations collecting and utilising data, the primary driver for data collection is often the pursuit of interests that can infringe on other people's data rights. Also, from a broader perspective, there is a lag in developing relevant legislation alongside weak enforcement mechanisms to oversee and regulate data-related practices. Addressing these root causes is crucial to ensuring the responsible and ethical handling of data across various industries and applications.

Similarly, Arroyo et al. (2021) observed that emerging technology like AI has continued to gain prominence in many industries, including construction, with its usage cutting across planning and control, generative design, and environmental performance assessments. However, the use of AI poses some ethical and social dilemmas which, if not properly addressed, can lead to the deployment of inappropriate solutions to solving inherent problems. In the same vein, Liang et al. (2024) noted that emerging technologies like AI and robotics are revolutionising construction project delivery, and their usage is becoming prominent within the construction industry of diverse countries. However, there is a need for ethical considerations in the adoption of these technologies due to their 'potential negative impacts' on job security, safety, and privacy. Unfortunately, this need for ethical considerations has not received significant attention in existing academic discourse on the adoption and implementation of emerging technologies in the construction industry. From the perspective of building information modelling (BIM) and BDA, Xing (2021) noted that while these technologies offer immense benefits to construction, they bring about new challenges. This includes issues relating to BIM model creation and information sharing, BIM data quality and security problems, and redistribution of industrial profits derived from BIM applications. These problems reinforce the need for ethics of technology within the construction industry.

Based on the foregoing, it can be deduced that professional ethics and ethics of technology are strongly correlated. They both strive towards ensuring morals in human activities and the use of emerging technologies. However, despite the continuous clamour on the need for ethical considerations in the use of emerging technologies (Dubber et al., 2020; Torresen, 2018), there is still little attention given to the ethics of technology, particularly in the construction industry where ethical concerns have been a prominent issue in the past. As such, exploring ethics and ethical practices in the sustainable use of emerging digital technologies in the construction industry becomes timely and of high necessity.

Understanding Sustainability

In the quest to ensure ethics in sustainable construction digitalisation, it is pertinent to delineate the concept of sustainability. According to Hendricks (2016), all around the world, people are living beyond their means. This is evident in the high consumption of energy and excess greenhouse gas emissions along with the wasteful use of natural resources. As such, Romiguer (2011) earlier noted that the environmental movement of the 1960s, which was concerned with human activities having severe and negative impacts on the planet, gave birth to the concept of sustainability. The report of Meadows, Randers, and Behrens in 1972 was one of the key works that highlighted this sustainability thinking, while the Brundtland Report (1987) provided sustainable development as the solution to this growing environmental problem. The Brundtland Report (1987) defined sustainable development as 'development that meets the needs of the present without compromising the ability of future generations to meet their own needs' (p. 43).

To achieve sustainable development in any country, sustainability thinking in the delivery of construction services is crucial. This is because the construction industry is crucial to the development of any country (Aghimien et al., 2021). At the first International Conference on Sustainable Construction held in Tampa, Florida, USA, in 1994, the conference convener defined sustainable development from construction perspective as an approach towards delivering a healthy built environment through efficient use of natural resources as well as ecologically based principles (Hussin et al., 2013). Sustainable development is a dynamic process that enables everyone to realise their potential and improve their quality of life in ways that simultaneously protect and enhance the earth's life support systems. Aghimien et al. (2018) observed that sustainability focuses on improving the quality of life for all of the earth's citizens without increasing the use of natural resources beyond the capacity of the environment to supply them indefinitely. Yekinni et al. (2015) opined that sustainable development involves transforming natural resources into productive output and, in the process, creating a balance between economic progress and environmental conservation, given that both are imperative to our future survival. Lehrer (2001) averred that sustainable development is aimed at ensuring continuous improvement of the standard of living without reversible damage to resources needed to survive as a species. Looking at each element separately, Alabi (2012) concluded that 'sustainability' means the capacity for an activity to be carried on indefinitely into the future, given the amount and type of resources available, whereas 'development' implies that society will change or grow in the same way. Hence, sustainable development suggests change and improvement that is compatible with environmental, social, and other limits, both now and in the long-term future (Gray & Wiedemann, 1999).

Akbiyikli et al. (2009), in exploring sustainability from a construction perspective, indicated that sustainable construction could be seen as a pathway through which the construction industry can contribute towards sustainable development, bearing in mind the environmental, socio-economic, and cultural pillars of sustainability. Sustainable construction incorporates the basic themes of sustainable development (Chaharbaghi & Willis, 1999; Parkin, 2000), and it brings about environmental responsibility, social awareness, and economic profitability objectives to the key players in the built environment and facilities for the wider community (Raynsford, 2000). Thus, sustainability in construction can be said to be a way of trying to find a balance between economic, environmental, and social factors in the design, construction, use, and maintenance of construction products and services.

Sustainability itself is a complex phenomenon which covers a wide range of topics, from habitat conservation and energy consumption to stakeholder satisfaction and financial results (Zhou et al., 2013). Considering that these topics affect construction, it is therefore necessary for the construction industry to address certain consequential issues in the process of achieving sustainable development. Moreover, the industry consumes a considerable amount of natural and physical resources and, as such, has a significant impact on the environment (Hendricks, 2016). Current building design, engineering methods, manufacturing technologies, and construction techniques need to be altered to accommodate requirements for sustainable construction.

Sustainability in Construction: The Triple Bottom Line

Studies have noted that sustainable development is rooted in the three pillars of social, economic, and environmental sustainability. To further emphasise this, John Elkington, in the mid-1990s, developed the Triple Bottom Line (TBL) framework to measure sustainability performance in corporate America. TBL was developed as an accounting framework, and it went beyond the traditional measures of profits, return on investment, and shareholder value to include environmental and social dimensions, as seen in Fig. 2.1. TBL inculcates the three dimensions of social, economic/finance, and environment, making it an effective tool in measuring the sustainability performance of construction projects. TBL is also commonly called the three Ps: People, Planet, and Profit, applied in the construction sector in the UK when the government sought to apply sustainable development practices to the construction industry (Slaper & Hall, 2011).

The Queensland Department of Public Works (2008) submitted that sustainability is a broad, all-encompassing concept that has implications for every aspect of society, including people of all ages and abilities. However, the concept does not

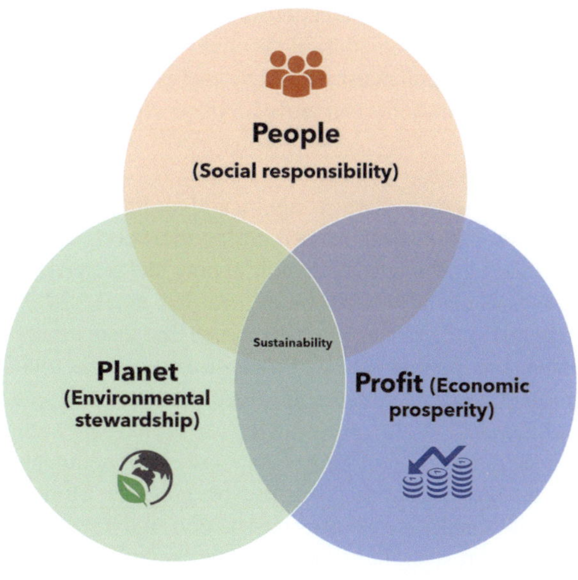

Fig. 2.1 The triple bottom line approach to sustainability. Source: Author's compilation (2024)

need to be difficult if viewed through the lens of TBL. The idea behind TBL reporting is sustainable success, which will benefit organisations as it provides a better reputation, increased confidence, benchmarked performance, increased operational efficiency, stakeholder satisfaction, and improved risk management. From the perspective of construction, it has been noted that sustainable construction is essential if optimum use of resources and minimal environmental impact is to be achieved. TBL offers construction organisations the opportunity to effectively measure the social, economic, and environmental impact of their projects and allows actionable insights for these organisations to ensure they deliver construction works sustainably (Consistent, 2023). Earlier, Awodele (2012) observed that each dimension of the TBL can be seen as (1) people who cut across the community and social impact, health and safety issues, fair compensation, equal opportunity, education, and recognition. (2) Profit, which entails revenue growth, productivity, profit margin, return on investment, capital cost, risk management, valuation of enterprise, and appeal to investors. (3) Planet includes energy, water and material use, emissions and waste, operational and design efficiencies, and new product and service opportunities.

Profit The profit dimension of businesses has been the major yardstick in measuring business prosperity. This aspect of the TBL reporting means measuring construction project performance in terms of the economic benefits they bring. The designs and construction, which are more cost-efficient in the long run, considering the selection of low maintenance materials, can lead to this needed economic prosperity. Extra expenses during construction can be provided for longer term operating cost savings. Determining the choice of design and material early in the project will eliminate the need to make additional costly changes to the construction later. Besides, the installation of smart energy and water features can also increase the value of the property. During the selection of building materials, the cost can be optimised through reduced transportation expenses by selecting materials from the local market. The use of appliances with high energy-star ratings that reduce the ongoing costs of running the home can prove highly useful economically (Aghimien et al., 2018).

Planet Construction projects interfere significantly with existing ecosystems due to the energy and raw materials consumed during their design, planning, erection, use, and disposal. They generate considerable energy and material flows along with impacts on local and global environments (Vasilca et al., 2021; World Green Building Council, 2019). Hussin et al. (2013) noted that many issues of sustainability are interrelated, and the interaction of a construction project with its surroundings has a significant impact on mankind. According to Constructing Excellence (2008) as well as Edwards (2001), reducing the use of non-renewable materials and water, as well as the production of emissions, waste, and pollutants, are some key sustainability challenges facing the construction industry. More than half of all resources consumed globally are used in construction, and 45% of the energy generated worldwide is used to heat, light, and ventilate the buildings, with a further 5%

arising from constructing them. The industry further contributes to 39% of global carbon emissions from the production of materials, construction and use of its products (International Energy Agency, 2019; United Nations Environment Programme, 2020; Vasilca et al., 2021; Weniger et al., 2023; World Green Building Council, 2019). Ultimately, while the construction industry and its activities are impacted by recent changes in climatic conditions, the industry, through excess energy consumption and carbon emission, has continued to contribute to the unsustainable built environment in countries around the world. Bossink and Brouwers (1996) suggested that the construction materials wasted on the site are relatively high. Water supply is also a growing cause for concern for the construction sector, which has particularly high requirements, especially in the manufacture of materials such as steel and concrete.

Therefore, from the planet perspective of the TBL, sustainable construction should minimise energy consumption, use non-renewable resources, and impact the environment through the optimum choice of construction components and energy resources (Guideline for sustainable building, 2016). Sustainable construction is resource-efficient because it incorporates sustainable and efficient management of water, waste, and energy. Also, using durable, non-toxic building materials during construction is important. Integrating renewable energy strategies, including passive solar design, is necessary, while conserving and protecting natural areas is of high necessity. The use of materials and products derived from sustainable-yield processes and/or manufactured locally should also be encouraged (Aghimien et al., 2018).

People Regarding the people's dimension of the TBL, Guideline for Sustainable Building (2016) submitted that all factors that influence the social and cultural identity of the people and how they assess their environment must be measured for effective sustainability. This is important in ascertaining how users and society assess a building. This means that a high degree of user satisfaction positively impacts the building's sustainability and, therefore, leads to a high level of acceptance and long-term value of the building. As such, all social aspects must be geared towards the individual and aim to create high-use value. This means that along with user health, comfort and satisfaction, appropriate building usability, and a high degree of spatial functional comfort are important factors. For a socially sustainable construction, the design must consider access, safety, and security. Moreover, it should cater to the users' future needs throughout their lives, such as young families, older residents, or residents with varying disabilities. It must be safe, easily adaptable to suit a diverse range of needs and comfortable for people with varying abilities at different stages of their lives. This is especially relevant considering the elderly among our ageing population. It is essential to note that an aesthetically pleasing and stimulating built environment will reinforce the sense of well-being of residents and people in the local community of the building (Aghimien et al., 2018).

Summary

The chapter strives to create an understanding of the concepts of digitalisation, ethics, and sustainability as they relate to the objective of this book. Industry 4.0 represents a significant transformation in manufacturing and production, building upon earlier industrial revolutions. This industrial revolution integrates emerging technologies like the IoT, AI, BDA, and robotics to enhance efficiency and innovation across various sectors, including construction, allowing for smart systems that adapt independently. A key emergence from this industrial revolution is the digitalisation of construction activities. This digitalisation entails using digital technologies to transform existing business models and generate new income streams while producing value-generating opportunities. However, Industry 4.0 also presents challenges, notably in the job market, where automation may displace some roles while creating new opportunities. This shift necessitates a re-evaluation of education systems. Additionally, rapid technological changes raise ethical and regulatory concerns, requiring updated frameworks to address privacy and accountability.

Generally, ethics covers moral principles that guide human behaviour, influencing decisions across various professions. It entails understanding right and wrong in professional activities, and in the construction industry, where unethical practices like corruption and mismanagement are prevalent, establishing a strong ethical framework is crucial for maintaining integrity and public trust. From the digitalisation perspective, the rise of emerging digital technologies necessitates a focus on both professional ethics and the ethics of technology. The difference between these two concepts lies in the fact that the ethics of technology address practical issues linked to technological advancements, including fairness, privacy, and security. As technology evolves, ethical considerations in its application must not be overlooked, particularly in construction, where the potential for negative impacts on job security and safety exists. Unfortunately, despite the importance of ethical frameworks, the construction industry has often neglected these discussions. To ensure the responsible use of emerging technologies, prioritising ethical standards that promote human well-being and social impact, enhancing project success and stakeholder confidence, has become quintessential.

From the sustainability perspective, the global issue of overconsumption and excessive greenhouse gas emissions has necessitated the global discourse on sustainable development. The construction industry plays a vital role in achieving sustainable development, requiring a focus on sustainability in construction services. A construction deemed sustainable should integrate environmental, socio-economic, and cultural considerations, aiming to balance these factors throughout a project's life cycle. As such, for as sustained construction digitalisation entailing the use of digital technologies in a way that promotes social responsibility, economic prosperity, and environmental stewardship, ethical use of these emerging technologies is essential. The next chapter of the book explores digitalisation in construction.

References

Aghimien, D. O., Aigbavboa, C. O., & Matabane, K. (2023). Dynamic capabilities for construction organisations in the fourth industrial revolution era. *International Journal of Construction Management, 23*(5), 855–864.

Aghimien, D. O., Aigbavboa, C. O., Oke, A. E., & Thwala, W. D. (2021). *Construction digitalisation—A capability maturity model for construction organisations*. Routledge.

Aghimien, D. O., Awodele, O. A., Oke, A. E., & Aghimien, E. I. (2018). Stakeholders' perception of sustainability in educational buildings in Nigeria. *International Journal of Sustainable Construction Engineering & Technology, 9*(1), 1–13.

Akbiyikli, R., Dikmen, S.U., & Eaton, D. (2009). Sustainability and the Turkish construction cluster: A general overview. In: *Proceedings of the construction and building research conference of the Royal Institution of chartered surveyors*, University of Cape Town, 10–11 September.

Alabi, A. A. (2012). Comparative study of environmental sustainability in building construction in Nigeria and Malaysia. *Journal of Emerging Trends in Economics and Management Sciences, 3*(6), 951–961.

Allen, G., & Chan, T. (2017). *Artificial intelligence and national security*. Belfer Centre for Science and International Affairs, Harvard Kennedy School. Retrieved December 11, 2023, from https://www.belfercenter.org/sites/default/files/files/publication/AI%20NatSec%20-%20final.pdf

Ameyaw, E. E., Pärn, E., Chan, A. P. C., Owusu-Manu, D., Edward, D. J., & Darko, A. (2017). Corrupt practices in the construction industry: Survey of Ghanaian experience. *Journal of Management in Engineering, 33*(6), 1–11.

Anjos-Santos, L. M., El Kadri, M. S., Gamero, R., & Gimenez, T. (2016). Developing English language teachers' professional capacities through digital and media literacies: A Brazilian perspective. In A. Manzoor (Ed.), *Handbook of research on media literacy in the digital age*. IGI Publishers.

Anton, E., Kus, K., & Teuteberg, F. (2021). Is ethics really such a big deal? The influence of perceived usefulness of AI-based surveillance technology on ethical decision-making in scenarios of public surveillance. In *Proceedings of the annual Hawaii international conference on system sciences* (pp. 2121–2130). HICSS.

Arroyo, P., Schöttle, A., & Christensen, R. (2021). The ethical and social dilemma of AI uses in the construction industry. In L. F. Alarcon & V. A. González (Eds.), *Proc. 29th annual conference of the International Group for Lean Construction (IGLC29)* (pp. 227–236). IGLC.

Ashok, M., Madan, R., Joha, R., & Sivarajah, U. (2022). Ethical framework for artificial intelligence and digital technologies. *International Journal of Information Management, 62*, 1–17.

Ashton, K. (2009). That internet of things thing: In the real-world things matter more than ideas. *RFID Journal, 22*, 97–114.

Awodele, O. A. (2012). *Framework for managing risk in privately financed market projects in Nigeria*. Heriot-Watt University School of the Built Environment.

Bahrin, M. A. K., Othman, M. F., Nor, N. H., & Azli, M. F. T. (2016). Industry 4.0: A review on industrial automation and robotics. *Jurnal Teknologi (Sciences and Engineering), 78*, 137–143.

Bédard-Maltais, P. (2017). *Industry 4.0: The new industrial revolution—Are Canadian manufacturers ready? BDC study*. Retrieved December 11, 2023, from www.bdc.ca/EN/Documents/analysis/bdc-resume-etude-manufacturing-en

Bienhaus, F., & Haddud, A. (2018). Procurement 4.0: Factors influencing the digitisation of procurement and supply chains. *Business Process Management Journal, 24*(4), 965–984.

Bossink, A. G., & Brouwers, H. J. H. (1996). Construction waste: Quantification and source evaluation. *Journal of Construction Engineering and Management, 122*, 55–60.

Bowen, P., Pearl, R., & Akintoye, A. (2007). Professional ethics in the south African construction industry. *Building Research and Information, 35*(2), 189–205.

Brennen, S., & Kreiss, D. (2014). *Digitalisation and digitisation*. Culture Digitally. Retrieved November 21, 2020, from http://culturedigitally.org/2014/09/digitalization-and-digitization/

Brundtland Report. (1987). Report of the World Commission on Environment and Development: Our common future. : United Nations General Assembly, Development and International Co-operation: Environment.

Brynjolfsson, E., & McAfee, A. (2014). *The second machine age: Work, Progress, and prosperity in a time of brilliant technologies*. W.W. Norton and Company.

Cartlidge, D. (2011). *New aspects of quantity surveying practice*. Spon Press.

Chaharbaghi, K., & Willis, R. (1999). Study and practice of sustainable development. *Engineering Management Journal, 9*(1), 41–48.

Chartered Institute of Building. (2018). *Rules and regulations of professional competence and conduct*. Retrieved June 30, 2024, from www.ciob.org/sites/default/files/2021-06/Rules%20and%20Regulation%20of%20Professional%20Competence%20and%20Conduct.pdf

Consistent. (2023). *Unlocking sustainability: The future of construction with sustainable buildings and the triple bottom line*. Retrieved August 12, 2024, from https://consistentconsultants.com/sustainable-buildings-and-the-triple-bottom-line/#:~:text=The%20TBL%20dimensions%20are%20commonly,environmental%20impact%20of%20the%20building

Constructing Excellence. (2008). *Key sustainability issues*. Retrieved March 5, 2015, from http://www.constructingexcellence.org.uk/zones/sustainabilityzone/sust_con/issues.jsp

Crampton, N. (2016). *SWOT analysis of the construction sector in South Africa*. Retrieved August 15, 2020, from https://bizconnect.standardbank.co.za/sector-news/construction/swot-analysis-of-the-construction-sector-in-south-africa.aspx

Dall'Omo, S. (2017). *Driving African development through smarter technology*. African Digitalisation Maturity Report (pp. 1–45).

Dimick, S. (2014). *Adopting digital technologies: The path for SMEs* (pp. 1–13). The Conference Board of Canada.

Dubber, M. D., Pasquale, F., & Das, S. (2020). *The Oxford handbook of ethics of AI*. Oxford University Press.

Edwards, B. (2001). *Rough guide to sustainability*. RIBA Publications.

Egbu, C., & Ofori, G. (2018). *Proceedings of the International Conference on Professionalism and Ethics in Construction*, Keyworth Centre, London South Bank University, 21–22 November 2018.

Erol, S., Jäger, A., Hold, P., Ott, K., & Sihn, W. (2016). Tangible industry 4.0: A scenario-based approach to learning for the future of production. *Procedia CIRP, 54*, 13–18.

Frey, C. B., & Osborne, M. A. (2017). The future of employment: How susceptible are jobs to computerisation? *Technological Forecasting and Social Change, 114*, 254–280.

Gartner. (2016). *Digitalisation*. Retrieved December 11, 2023, from https://www.gartner.com/en/information-technology/glossary/digitalisation

Gibson, I., Rosen, D. W., & Stucker, B. (2015). *Additive manufacturing technologies: 3D printing, rapid prototyping, and direct digital manufacturing* (2nd ed.). Springer.

Gray, P. C. R., & Wiedemann, P. M. (1999). Risk management and sustainable development: Multinational lessons from approaches to the use of indicators. *Journal of Risk Research, 2*(3), 201–218.

Green, B. P. (2024). What is technology ethics? Retrieved August 12, 2024, from https://www.scu.edu/ethics/focus-areas/technology-ethics/#:~:text=Technology%20ethics%20is%20the%20application,t%20have%20to%20make%20before

Grieves, M., & Vickers, J. (2017). Digital twin: Mitigating unpredictable, undesirable emergent behaviour in complex systems. In *Transdisciplinary Perspectives on Complex Systems* (pp. 85–113). Springer.

Guideline for Sustainable Building. (2016). *Future-proof design, construction and operations of buildings*. Federal Ministry for the Environment, Nature Conservation, Building and Nuclear Safety.

Hendricks, B. (2016). *Guideline for sustainable building: Future-proof design, construction and operations of buildings*. Federal Ministry for the Environment, Nature Conservation, Building and Nuclear Safety.

Hussin, J. M., Rahman, I. A., & Memon, A. H. (2013). The way forward in sustainable construction: Issues and challenges. *International Journal of Advances in Applied Sciences, 2*(1), 15–24.

International Energy Agency. (2019). *Global status report for buildings and construction 2019*. International Energy Agency. Retrieved August 8, 2023, from https://shorturl.at/dmzGZ

Irniger, A. (2017). *Difference between digitisation, digitalisation and digital transformation*. Retrieved August 12, 2024, from https://www.coresystems.net/blog/difference-between-digitization-digitalization-and-digital-transformation

Johnstone, K., Kervin, L., & Wyeth, P. (2024). *Defining digital technology*. Retrieved August 12, 2024, from https://digitalchild.org.au/defining-digital-technology/

Kalavendi, R. (2017). *Digitisation and digitalisation: The two-letter difference*. Retrieved November 19, 2020, from https://medium.com/@ravisierraatlantic/digitization-digitalization-the-2-letter-difference-59b747d42ade

Khodabakhshian, A. (2024). Ethics-aware application of digital technologies in the construction industry. *Springer Briefs in Applied Sciences and Technology, 2336*, 49–64.

Lasi, H., Fettke, P., Kemper, H.-G., Feld, T., & Hoffmann, M. (2014). Industry 4.0. *Business and Information Systems Engineering, 6*(4), 239–242.

Lehrer, D. (2001). Sustainable design. In E. Teicholz (Ed.), *Facility design and management handbook*. McGraw-Hill.

Leviakangas, P., Paik, S., & Moon, S. (2017). Keeping up with the pace of digitisation: The case of the Australian construction industry. *Technology in Society, 50*, 33–43.

Liang, C., Le, T., Ham, Y., Mantha, B. R. K., Cheng, M. H., & Lin, J. J. (2024). Ethics of artificial intelligence and robotics in the architecture, engineering, and construction industry. *Automation in Construction, 162*, 105369.

Lin, H., & Xu, Q. (2021). Research on data ethics based on big data technology business application. *Advanced Intelligent Systems Computing, 1303*, 1391–1396.

Liu, Z., & Wang, G. (2020). Ethics of technology and engineers' professional ethics. *Philosophy of Engineering and Technology, 34*, 57–64. https://doi.org/10.1007/978-981-15-1952-9_5

Longstaff, S. (1999). The good, the bad and the ugly: Some thoughts on the ethical dimension of architecture. *Architecture Australia, 13*, 4–8.

Lu, Y. (2017). Industry 4.0: A survey on technologies, applications and open research issues. *Journal of Industrial Information Integration, 6*, 1–10.

Mansfield, J. R. (2008). The ethics of conservation: Some dilemmas in cultural built heritage projects in England. *Engineering Construction and Architectural Management, 15*(3), 270–281.

Mayer-Schönberger, V., & Cukier, K. (2013). *Big data: A revolution that will transform how we live, work, and think*. John Murray Publishers.

McAleenan, P., McAleenan, C., Ayers, G., Behm, M., & Beachem, Z. (2019). The ethics deficit in occupational safety and health monitoring technologies. *Proceedings of Institution of Civil Engineers: Management, Procurement and Law, 172*(3), 93–100.

Monton, A. L. (2022). *Difference and similarities: Digitisation, digitalisation, and digital transformation*. Retrieved August 12, 2024, from https://www.globalsign.com/en-sg/blog/difference-and-similarities-digitization-digitalization-and-digital-transformation#:~:text=While%20digitization%20focuses%20on%20converting,generate%20insights%20from%20their%20behaviour

Ochs, T., & Riemann, U. T. (2018). IT strategy follows digitalisation. In *Encyclopedia of information science and technology* (4th ed.). IGI Publishers.

Oladinrin, O. T., Arif, M., Rana, M. Q., & Gyoh, L. (2023). Interrelations between construction ethics and innovation: A bibliometric analysis using VOSviewer. *Construction Innovation, 23*(3), 505–523.

Oladinrin, T. O., & Ho, C. M. F. (2014). Strategies for improving codes of ethics implementation in construction organisations. *Project Management Journal, 45*(5), 15–26.

Olatunji, S. O., Oke, A. E., Aghimien, D. O., & Ogunwoye, O. S. (2016). Implementation of code of ethics among quantity surveying firms in Nigeria. *Developing Country Studies, 6*(5), 71–76.

Parkin, S. (2000). Context and drivers for operationalising sustainable development. *Proceedings of ICE, 138*, 9–15.

Paul, C. A., Aghimien, D. O., Ibrahim, A. D., & Ibrahim, Y. M. (2021). Measures for curbing unethical practices among construction industry professionals: Quantity surveyors' perspective. *Construction Economics and Building, 21*(2), 1–17.

Pullen, D. L. (2009). Back to basics: Electronic collaboration in the education sector. In J. Salmons & L. Wilson (Eds.), *Handbook of research on electronic collaboration and organizational synergy*. IGI Publishers, Pennsylvania.

Queensland Department of Public Works. (2008). *Smart and sustainable homes design objective*. The State of Queensland (Department of Public Works). Retrieved December 11, 2023, from www.smarthousing.qld.gov.au

Raynsford, N. (2000). Sustainable construction: The government's role. *Proceedings of ICE, 138*, 16–22.

Robinson, S., Dixon, J. R., Preece, C. N., & Moodley, K. (2007). *Engineering, business and professional ethics*. Butterworth-Heinemann.

Romiguer, A. T. (2011). *Sustainable development: Objectives, enablers and challenges for Spanish companies*. Tampere University of Technology.

Royal Institute of Chartered Surveyors. (2021). *Rules of conduct*. Retrieved June 30, 2024, from https://www.rics.org/content/dam/ricsglobal/documents/standards/2021_roc_en.pdf

Russell, S. J., & Norvig, P. (2016). *Artificial intelligence: A modern approach*. Pearson Education Limited.

Rüßmann, M., Lorenz, M., Gerbert, P., & Waldner, M. (2015). Industry 4.0: The future of productivity and growth in manufacturing industries. *Boston consulting group*, 2015 Report.

Schwab, K. (2017). *The fourth industrial revolution* (1st ed.). Crown Business.

Slaper, T. F., & Hall, T. J. (2011). The triple bottom line: What is it and how does it work? *Indiana Business Review, 86*, 4–8.

Smowltech. (2024). *Digital ethics: Definition and examples*. Retrieved August 12, 2024, from https://smowl.net/en/blog/digital-ethics/#:~:text=Digital%20ethics%20refers%20to%20the,and%20society%20as%20a%20whole

Society of Chartered Surveyors (Ed.). (2006). *Assessment of professional competence: Candidates guide* (2nd ed.). Society of Chartered Surveyors.

Sung, T. K. (2018). Industry 4.0: A Korea perspective. *Technological Forecasting and Social Change, 132*(C), 40–45.

Torresen, T. (2018). A review of future and ethical perspectives of robotics and AI. *Frontiers in Robotics and AI, 4*, 75. https://doi.org/10.3389/frobt.2017.00075

United Nations Environment Programme. (2020). *2020 global status Report for buildings and construction: Towards a zero-emission, efficient and resilient buildings and construction Sector*. Global Status Report. Retrieved August 12, 2024, from www.iea.org

Vasilca, I.-S., Nen, M., Chivu, O., Radu, V., Simion, C.-P., & Marinescu, N. (2021). The management of environmental resources in the construction sector: An empirical model. *Energies, 14*(9), 1–19.

Verhulst, S. (2002). About scarcities and intermediaries: The regulatory paradigm shift of digital content reviewed. In L. A. Lievrouw & S. Livingstone (Eds.), *The handbook of new media* (pp. 432–447). Sage Publications.

Weniger, A., Del Rosario, P., Backes, J. G., & Traverso, M. (2023). Consumer behavior and sustainability in the construction industry—Relevance of sustainability-related criteria in purchasing decisions. *Buildings, 13*(3), 638.

World Green Building Council. (2019). Bringing embodied carbon upfront. Retrieved August 12, 2024, from https://worldgbc.org/advancing-net-zero/embodied-carbon/

Xing, J. P. (2021). Engineering ethics of BIM and big data application in civil engineering. *E3S Web of Conferences, 248*, 1–4.

Yekinni, A. A., Bello, S. K., & Olaiya, K. A. (2015). Application of value engineering techniques in sustainable product and service design. *Science and Engineering Perspectives, 10*, 120–130.

Zhou, L., Keivani, R., & Kurul, E. (2013). Sustainability performance measurement framework for PFI projects in the UK. *Journal of Financial Management of Property and Construction, 18*, 232–250.

Chapter 3
Digitalisation in the Construction Industry

Introduction

The construction industry has been described as a crucial contributor to socio-economic development through its contribution to gross domestic product (GDP), employment, and infrastructure development. It is one of the important sectors that play a role in the economy (Oladinrin et al., 2012). World Economic Forum (2016) also submitted that the construction industry globally is critical to societal, economic and environmental growth. This is important as many businesses across diverse sectors rely on the construction industry to develop and maintain their infrastructure. Despite this important nature of the construction industry, the industry performance has been a topical issue. Poor performance of project delivery has been noted, and several issues facing the industry have been identified in existing studies (Agyekum et al., 2022; Aigbavboa et al., 2022; Arndt et al., 2020; Heimgartner, 2020). Among these issues is the slow adoption of emerging digital technologies, which has resulted in a lack of innovation in many construction industries globally (Alsharef et al., 2024). Advancements in technologies have continued to revolutionise how many industries operate, and construction is no exception. Emerging digital technologies offer solutions to the construction industry's poor performance. Therefore, this chapter on digitalisation in the construction industry is designed to explore the structure of the global construction industry and the challenges facing the use of emerging digital technologies within the industry. Furthermore, diverse digital technologies were explored to give construction organisations, industry participants, and stakeholders insights into the available technologies and their benefits. The chapter further explores the deployment of digital technologies in the construction industry of developed and developing countries with a view to unearthing the extent of usage, the challenges being faced, and the measures needed to improve the use.

O. Oladinrin et al., *Ethical Practices for Sustainable Construction Digitalisation*, https://doi.org/10.1007/978-3-032-00151-1_3

The Construction Industry

The term 'construction' is the process of creating and repairing immovable structures. It has also been described as the process where the services of experts known as 'contractors' are engaged to deliver structures for defined purposes. These structures can be in the form of buildings designed for private or public use or public infrastructure such as roads, dams, bridges, and railways, among others (Nuzha, 2023). The construction industry is responsible for ensuring the construction projects are delivered to meet the needs for which they have been designed. The industry, through the physical implementation of various construction projects, has become crucial in transforming people's aspirations into reality (Ibrahim et al., 2010). As such, the construction industry globally has come to be known as a driver of infrastructure growth as well as the socio-economic development of many countries (Ofori, 2018). Bohari et al. (2015) noted that the construction industry contributes to economic growth by providing tangible services and infrastructure in societies. Ayodele and Alabi (2011) averred that the wellness of the economy of any country can be evidence of the growth of the country's construction industry. This can be a result of the fact that many other industries that also add to economic growth depend on the product of the construction industry to function. In fact, Stasiak-Betlejewska and Potkány (2015) have earlier noted that the construction industry is linked to other parts of economic activity such as manufacturing and energy usage, materials, finance, labour, and equipment. As such, the activities of the construction industry can directly or indirectly impact other industries, ultimately affecting the country's economic growth (Skudder, 2021).

According to the World Economic Forum (2016), the construction industry contributes up to 6% of the world's GDP. In 2018, a significant increase of up to 13% in global GDP contribution worth about 10 trillion dollars was recorded, with a projected growth of up to 14 million dollars in 2025 (Ryfan, 2018). This important contribution of the industry to economic development is evident in diverse countries. The Chartered Institute of Building (2018) noted that the construction industry contributed 6% to the UK's GDP in 2016 and has been a significant driver of employment, productivity and business growth in the country. However, Mitchell (2020) noted that as of 2018, the contribution of the industry to economic growth had increased to 8%. In the United States of America, a 4.2% contribution was recorded in 2021 (Associated General Contractors of America, 2022). In Australia, the construction industry contributes up to 10% to the GDP, making it the fifth largest sector in the country (International Trade Administration, 2024). Across Europe, a 9% contribution to the GDP of diverse countries has been noted (The European Commission, 2020). In China, the industry contributes about 6.8% to the GDP (Statista Research Department, 2019), while an 8% contribution has been noted in India (Dixit et al., 2017). The Department of Statistics Malaysia (2020) noted a 4.2% contribution to the Malaysian GDP, while 3.21% and 3% contributions have been noted in Nigeria and South Africa, respectively (National Bureau of Statistics, 2020; Zingoni, 2020). In Kenya, a contribution of 4.8% was recorded, which is expected to grow in subsequent years (Wetangula & Mazurewicz, 2017). These diverse submissions show the importance of the construction industry to the growth of countries worldwide.

Regarding employment provision, Durdyev and Ismail (2012) noted that the construction industry is one of the highest employers of labour, with over 111 million employed globally. Recent submissions show that the industry accounts for about 8.6% of global employment (My Job Quote, 2024). In the UK, it was noted that the construction industry contributes over 10% to job creation (Mitchell, 2020). In the US, 4.8% of the entire workforce can be found in the construction industry (TST Europe, 2023). The story is no different in China, where over 51 million people were employed within the construction industry as of 2022 (Zhang, 2024). The Australian Government (2024) also noted that 9.4% of its working class have jobs in the construction industry. Also, the construction industry has made a 3% contribution to employment in Ghana, Nigeria and Zambia, with 11% in South Africa (Ahadzie, 2019; Chartered Institute of Building, 2018; Cheelo & Liebenthal, 2018; Oladinrin et al., 2012). These employment statistics affirm Dithebe et al. (2018) earlier submission that the construction industry worldwide can bridge societal inequality by providing jobs to millions of workers. These jobs range from management to specialised teams, skilled, partly skilled, and under-skilled workforce (Vitharana et al., 2015).

In delivering construction projects, the industry requires various project participants whose actions significantly impact the outcome of construction projects. Bikitsha et al. (2014) described construction projects as corporate ventures that include diverse professionals brought together to create a construction project team and to accomplish the shared objective of delivering a successful project. Construction projects encompass short-term management with self-directed expertise that depends on the using of resources received from the client and firms. Buvik and Rolfsen (2015) noted that it is a collaborative endeavour that involves several distinct organisations brought together to establish a project team. The project team collaborates to accomplish the shared goals of successfully delivering the construction project, which involves several activities.

This dependence on diverse professionals and groups makes the construction industry complex, with multifaceted activities designed to meet the construction clients' desires. Navon (2005) noted that the complex nature of the industry stems from the involvement of many stakeholders, such as clients, contractors, consultants, stakeholders, financial institutions, and regulators, among others. These large number of project participants, coupled with clients' growing demand, the complexity of construction projects, advancement in technology and introduction of innovations, amongst others, have made the construction industry a dynamic and demanding industry when compared to others (Aghimien et al., 2021). Based on this demanding nature of the industry, Takim and Akintoye (2002) posited that the level of success in construction activities will depend heavily on the quality of the managerial, financial, technical and organisational performance of the diverse project participants while taking into consideration the associated risk management, the business environment, and economic and political stability.

Issues Facing the Construction Industry

Globally, the construction industry is challenged by several issues. The coronavirus (COVID-19) pandemic brought its share of global issues to the construction industry (Aigbavboa et al., 2022). Office of National Statistics (2020) noted a shrink in the UK's construction industry output in the early period of the nationwide lockdown, which led to a significant loss of jobs. Several problems revolving around labour and material shortages were prevalent with in UK's construction industry. Also, in the US, Heimgartner (2020) noted the stoppage of construction activities, leading to a supply chain disruption and loss of jobs for construction workers. In Ghana, delayed payment, perverse rise in the cost of construction materials and slow rate of completion of construction projects were prevalent (Agyekum et al., 2022). In South Africa, the construction industry recorded up to a 60% reduction in output, leading to job loss for construction organisations and individuals (Aigbavboa et al., 2022; Arndt et al., 2020). In Malaysia, many construction projects were discontinued, while delays in delivering others where common (Gamil & Alhagar, 2020). The story is no different in Oman, where projects were delayed, and many materials were unavailable to deliver projects. Also, shortage of staff and economic decline were some of the issues faced by the construction industry (Amri & Marey-Pérez, 2020).

Before COVID-19, Leeds (2017) observed that the continuous population growth poses a problem for the construction industry. Given that the construction industry is responsible for the delivery of infrastructure, it is expected that as the population increases, the demand for infrastructure also increases, thereby exerting pressure on the construction industry. Alshalalda (2019) noted that the population expands by approximately 83 million annually. Evidently, this rise in urban population necessitates a reappraisal of the construction industry to boost productivity while providing affordable buildings and advancing infrastructure capacity (Buehler & Gerbert, 2017). Achieving this feat poses challenges to construction organisations and participants. According to Leeds (2017), issues such as poor productivity and profitability, poor project performance, shortage of skilled workers, and concerns around sustainability were highlighted as key challenges faced by the industry worldwide.

Eventura (2023) highlighted issues such as disruption in the construction supply chain, the rising cost of energy, shortage of labour, rising interest rates and inflation, excessive emission of carbon and poor sustainability thinking, and economic crisis leading to slow investment from private clients. Also, Barton (2023) submitted that the construction industry grapples with issues of slow growth rate, ageing workforce and shortage of skilled workers, supply chain disruption, rising costs and inflation, climate change, sustainability problems and a lack of technology adoption. A similar observation was made in Lewis (2023), where issues around sustainability, transformation, integrating emerging technologies and project governance were observed as key challenges facing construction. Massay (2024) suggested that the main issues facing the global construction industry centres around unsustainable construction and excessive carbon emission, shortage of skills, failure to integrate technology, poor commination, cash flow and bad debt, health and safety risks, and gender imbalance. Blackridge (2024) gave a more extensive list of global issues facing the construction industry. These include the prevailing cost and time overrun of projects, slow adaptation to emerging technologies, poor communication on projects, shortage of workers, poor project planning, forecasting and budgeting, poor organisational management, cash flow problems and poor scheduling. Looking at these recent challenges facing the construction industry, it is evident that skills shortages, poor emerging technology adoption and sustainability considerations are common. This underscores the importance of this book, which is designed to ensure the ethical use of digital technologies for the delivery of sustainable construction.

Alsharef et al. (2024) submitted that the construction industry faces three major issues in the US. These are skill shortages, supply chain disruptions, and the slow integration of emerging technologies. Aghimien et al. (2021) have also noted that the issue of skill shortage has been an on-going challenge for the construction industry in many developing countries. Windapo (2016) described this skills shortage as an inadequate supply of well-qualified workers prepared to work under the current market conditions, especially at standard wages. This phenomenon is caused by rapid technological advancements, an ageing workforce, the role of the state, the quality and applicability of training by artisans, the industry's poor image, and economic conditions. Moreover, due to globalisation, the international construction market is accessible to firms in both developing and developed countries. This means that construction activities and personnel can be exported, with most coming from developed countries to developing countries (Ofori, 2000). Lewis (2007) emphasised the effects of globalisation on the construction industry of developing countries and the regions in which global trade preserves the economic deficit, therefore creating a challenge to developing the construction industry in developing countries. Ngowi and Lema (2002) have earlier noted that globalisation affects the sustainability of construction firms in developing countries. According to Khan (2004), to minimise the impacts of globalisation, developing countries must adapt particular policies including training labour to manage the challenges of an open economy, obtaining new technological skills, developing local industry and observing performance monitoring systems.

Aghimien et al. (2021) noted that the construction industry suffers from a poor image. The industry is considered dirty, difficult, and dangerous. International Labor Organisation (2001) noted that to many people, the construction industry is only for those interested in blue-collar jobs, which is mostly attributed to low-status and according to Tucker et al. (1999), for many people, a career in construction does not offer a clear career path. Moreover, Haupt and Harinarain (2016) noted that construction's dirty, difficult and dangerous nature makes the sector unattractive to younger people. This situation is worsened by the common unethical practices by which the construction industry in now known (Paul et al., 2021).

The issue of construction site safety has also been a common problem in the construction industry of many countries. Although considerable attention has been garnered by construction health and safety worldwide, the construction industry still contributes significantly to injuries and fatalities in the workplace. Although the cases are worse in developing and underdeveloped countries, the persistence of site accidents leading to severe injuries and fatalities in the construction industry globally, is alarming (Aghimien et al., 2021). The importance of attaining a safe working environment has necessitated diverse approaches being proposed, including the proper health and safety management at diverse phases of construction, the development of safety models and practices, safety training, and the use of technological innovations (Chan & Aghimien, 2022; Ibrahim et al., 2023; Ikuabe et al., 2021). These approaches became necessary due to the dire impact of construction site accidents on individuals and organisations. When accidents occur, there is always a significant loss of production time and compromise of construction quality. This is coupled with the associated pain and suffering of the victim's family, the reduction in staff morale as well as the high medical cost incurred (Kadiri et al., 2014). This poor health and safety in the industry results in a poor image that transcends to a skills shortage as potential workers are scared to venture into the industry (Aghimien et al., 2021).

Based on the aforementioned issues, the challenges facing the construction industry can be summarised into the following:

1. Cash flow and bad debt.
2. Climate change.
3. Continuous rise in urban population.
4. Excessive emission of carbon.
5. Gender imbalance.
6. Global economic crisis.
7. Globalisation.
8. Health and safety risks.
9. Poor commination.
10. Poor image of the industry.
11. Poor organisational management.
12. Poor productivity and profitability.
13. Poor project governance.
14. Poor project performance.
15. Poor project planning, forecasting and budgeting.
16. Poor scheduling.
17. Poor sustainability thinking.
18. Rising cost of energy.
19. Rising interest rates and inflation.
20. Shortage of skilled labour and ageing workforce.
21. Slow adaptation to emerging technologies.
22. Supply chain disruption.
23. Unethical practices.
24. Unwillingness to embrace change.

Digital Technologies in Construction

Technological advancements have continued to revolutionise how many industries operate, and construction is no exception. Despite technology adoption being a crucial issue facing the construction industry worldwide, sightings of the adoption of some emerging digital technologies have continued to emerge within the body of literature. Some of these technologies are discussed in this section.

Internet of Things

Internet of Things (IoT) refers to technologies (electric or electronic devices) connected by a smart network that contains knowledge control processes and automatically connects with other surrounding devices. The smart network is a connection that allows for remote monitoring and control over the object (Lee & Lee, 2015). Kumar et al. (2016) noted that IoT entails four layers: application, perception, network, and physical. The application layer pertains to the specific domains where the technology is applied, while the perception layer encompasses the technologies that facilitate communication with other objects. The network layer is responsible for network communication and coverage, and the physical layer consists of the hardware, such as smart appliances and other devices.

Shvets and Hanák (2022) noted that IoT is an important evolution in the context of construction 4.0. According to Rudyi (2024), IoT is essential in the transformation of industries, automating processes and ensuring return on investment. It was further noted that through IoT sensors and other technologies, organisations can improve the visibility on construction sites from inception to completion of a project. Newton (2023) highlighted the unique application of IoT in construction. It was noticed that IoT is mostly used in wearables, safety sensors, material monitoring, theft prevention and waste management. Rudyi (2024) also noted some uses of IoT in construction, and these include just-in-time provisioning using Radio Frequency Identification (RFID), tracking of tools and equipment using IoT-enabled sensors and GPS, remote operations and monitoring of activities, and saving cost on fuel consumption among others. Also, Shvets and Hanák (2022) explored the use of IoT in different project phases and noted that during the construction phase, IoT is mostly used in monitoring the quality of projects and ensuring health and safety. However, at the operational phase, IoT is most adopted to introduce smart concepts and energy management using of completed structures. BIM Engineering (2018) noted that the application of IoT in construction is mostly prevalent in the areas of preventive maintenance of buildings, real-time monitoring and observation using sensors, management of construction works through effective decision-making, human resource monitoring, and site safety.

Benefits of IoT are recorded as smart communication, remote control, sensors on machinery forecast when maintenance is required, and supply replenishment, i.e. when supply units are labelled with an RFID tag, the site system can monitor quantities and reorder materials (Mahmud et al., 2018). Also, Urie (2019) noted that IoT offers the construction industry the opportunity to deliver smart buildings through sensors and interconnected technologies. It also offers Just-in-time provision, easy tracking of tools and equipment, monitoring of workers' activities and health remotely, energy conservation and effective use of augmented reality.

Unmanned Aerial Vehicles

Unmanned Aerial Vehicles (UAVs), also known as drones in construction sites, are increasingly being used to monitor construction activities (Edirisinghe, 2018). UAVs are also called 'remotely piloted aircraft' because they can be operated without a pilot onboard due to their compact design and size (Chen et al., 2022; Golizadeh et al., 2019). According to Chen et al. (2022), UAVs are used to collect digital information through remote or autonomous operations. The use of UAVs has stimulated notable research interests in the construction industry. Some studies have examined the utilisation of UAVs to address diverse construction issues, including project monitoring, workers' safety, and risk management on construction sites (Ghasri & Maghrebi, 2021; Jeong et al., 2020; Martinez et al., 2020; Moud et al., 2021; Sreenath et al., 2020). Additionally, UAVs have been explored in terms of improving the management of projects and their integration with other digital technologies for efficient construction processes (Alladi et al., 2020; Liu et al., 2020; Zhou et al., 2018). Sreenath et al. (2020) noted that UAVs are easy to use and maintain. They are considered reasonably affordable and are capable of carrying cameras, visuals, or infrared and other sensors that can be used to capture digital data. Integrating the digital data gathered with Building Information Modelling (BIM) can aid in effective monitoring, reporting, and improving project delivery (Golizadeh et al., 2019). According to Elghaish et al. (2021), the increased use of UAVs can automate surveying, aerial photography, and inspection processes on construction sites. Numerous studies have also noted that the use of UAVs in construction includes improved supply chain, better quality management, improved security, and risk decision-making (Dupont et al., 2017; Goessens et al., 2018; Li & Liu, 2018; Martinez et al., 2020). Propeller Aero (2024) noted that construction organisations could employ UAVs when carrying out site analysis, designing, asset inventory, reporting and collaboration, as well as dispute resolutions. Using UAVs promises increased efficiency, reduced expenses, improved safety standards, and data accuracy to construction organisations that are able to implement it effectively, particularly in the area of surveying. It was further noted that compared to traditional surveying, UAVs offer cost savings and timely delivery of site surveys.

Building Information Modelling

Building Information Modelling (BIM) enables the creation of a digital simulation or blueprint of a construction project before the physical building begins. This digital model visually represents the building's physical and functional characteristics (Arayici & Aouad, 2010). BIM has been described as an IT-propelled tool that creates and maintains a digital representation of building processes and information throughout the project lifecycle in the form of a data repository. BIM incorporates 3D modelling and can interpret and apply intelligent relationships between all elements in a building model, including both geometric and non-geometric data (Chimhundu, 2016). From the perspective of Azhar (2011), BIM is not just a software application but also a process that facilitates changes in project delivery workflows, going beyond the use of 3D intelligent models.

Kiprotich (2014) noted that BIM is useful throughout a construction project lifecycle, from design to maintenance. The application of BIM starts in the pre-design stage, where information related to project work plans, data and documentation collection, site surveys and testing, and space design confirmation are recorded. The recorded data is then inputted into the BIM software, making it readily available to all project stakeholders. This increases the speed of information exchange and delivery between parties. Azhar (2011) noted earlier that since information models are created in 3D space, they can automatically be used to check for interferences, detect conflicts, and identify collisions across all major building systems. BIM can also be used to precisely measure quantities of materials and components, leading to efficient procurement processes throughout the design and construction stages (Ghaffarianhoseini et al., 2016). BIM is also useful for preparing estimates and schedules, managing and tracking changes, and managing site logistics. During the construction phase, BIM can be used for construction sequencing, arranging, ordering and delivering materials on schedule, and assisting in the fabrication of building components (Azhar, 2011; Kiprotich, 2014). BIM also enables a two-way flow of information, allowing it to be used as a manual to safely manage and operate the building, providing all necessary information about the facility's electrical, mechanical, and physical. Also, BIM models can replicate processes for maintenance and retrofitting, reducing costs related to facility management and enhancing the maintenance process (Ghaffarianhoseini et al., 2016).

BIM is a tool used in the construction industry to envision and coordinate industry activities, leading to improved productivity, reduced errors and omissions, enhanced quality management, and better scheduling, cost control, and safety on projects. Adopting BIM in construction projects offers several benefits, including improved quality of designs, enhanced speed and cost-effectiveness of professional services, reduced design errors through design interrogation, better financial project control, improved information sharing and synchronised work among multiple professionals (Bataw et al., 2016). Yet, many opportunities to integrate BIM with other digital technologies, such as using the 3D models created through BIM for 3D visualization or 3D printing, remain unrealised (Arayici & Aouad, 2010).

Big Data Analytics

Big data analytics (BDA) is a disruptive paradigm that inevitably impacts the construction industry. Unfortunately, unlike other industries, the construction industry has yet to fully capitalise on the potential of BDA to solve its prevalent issues and transform its operations (Ismail et al., 2018). The current prominence gained by BDA can be attributed to the fourth industrial revolution's advanced technologies, which have increased the rate of data generation (Waal-Montgomery, 2015). This surge in data and the large nature of these data necessitates interpretation, which is where BDA comes into play, leveraging the 3Vs of BDA: Volume, Velocity, and Variety (Ismail et al., 2018).

BDA entails searching for patterns and insights within large, diverse data sets known as 'Big Data' that are generated through Big Data engineering (Atzori et al., 2010). According to Gartner (2012), these Big Data are high-volume, high-velocity, and high-variety information assets that require innovative, cost-effective processing to enable enhanced insights, decision-making, and process automation. These data sets are too large and complex to be managed and processed by typical systems (Chahal & Gulia, 2016; Manyika et al., 2011). To extract value from Big Data, BDA is essential as it leverages technology to gather, process, and analyse large, real-time data streams to obtain insights that can drive market leadership (Cabrera-Sánchez & Villarejo-Ramos, 2019; McAfee & Brynjolfsson, 2012; Rüßmann et al., 2015). Essentially, BDA processes the complex, massive datasets known as Big Data (Nizam & Hassan, 2017). It encompasses various specialised techniques such as data mining, text mining, visual mining, process mining, and real-time data analysis (Atzori et al., 2010; Davenport & Patil, 2012). BDA draws from an extensive intellectual field such as data science, statistics, predictive analytics, and knowledge discovery (Datt, 2011). According to Chen and Zhang (2014), effective BDA requires proper data collection and configuration for reliable conclusions. Importantly, BDA highly depends on human skills and knowledge to leverage Big Data techniques and tools (Dubey, 2019; Gupta & George, 2016). According to Gupta and George (2016), human capabilities contribute up to 72% of the effectiveness of the BDA process.

The construction industry deals with a substantial amount of heterogeneous, underutilized data (Bilal et al., 2016; Burger, 2017). This data surge is driven by the industry's adoption of advanced technologies and IoT. Addo-Tenkorang and Helo (2016) suggest that this influx of data in construction presents new opportunities to extract valuable insights that can drive industry improvements if the unutilised data can be effectively leveraged. Also, Peiffer (2016) propose that the data generated in construction can enhance productivity if leveraged effectively through real-time data exchange. BDA can aid in transforming diverse data formats into actionable insights to improve construction operations.

Augmented and Virtual Realities

Augmented reality (AR) allows users to view a real-world scene with computer-generated information overlaid (Chi et al., 2013). In contrast, virtual reality (VR) creates a fully immersive 3D environment that surrounds the user (Li et al., 2018). AR and VR technologies have various applications in the construction industry. Site managers and workers can use AR on portable devices like tablets to effectively manage construction defects (Kwon et al., 2014). Additionally, VR-based safety training platforms have been developed to improve construction safety by enabling university students to access safety information using QR codes on mobile devices. VR can also be used for equipment and task training through simulation-based systems, such as a virtual training system for construction excavators (Wang et al., 2018). Furthermore, wearable devices have been created to provide engineers with on-site access to construction drawings and related information without having to physically bring the drawings to the site. Moreover, an AR planner system has been developed to help construction site planners virtually arrange materials and equipment using 3D models in the planned worksite (Kodeboyina & Varghese, 2016).

Robotics and Automation

The application of robotics and automation in construction has been linked to issues of labour shortage, an ageing workforce and high accidents and injury risks in the industry (Aghimien et al., 2021; Cai et al., 2018). Struková and Líška (2012) noted that the application of automation and robotics in construction is extensive, spanning the entire construction life cycle—from planning and design to execution, maintenance, monitoring, and even demolition. New robotic systems have been developed to automate various construction processes, such as bricklaying, internal finishes, and prefabricated building construction (Yahya et al., 2019). For civil works, automation is applied on-site and off. In the prefabrication of construction components, automation and robotics are widely used in pre-cast concrete production, enabling a shift from mass production to mass customisation. Robotic cells can carry out multiple tasks, such as placing reinforcement bars and setting moulds, facilitating this transition (Bock, 2008). Robots are more efficient in delivering safety for dangerous work as humans are away from harms-way (Aghimien et al., 2021; Bahrin et al., 2016). Chan et al. (2016) further submitted that BIM automation can enhance safety management's effectiveness and efficiency.

Cloud Computing

The Internet is the cornerstone of different technologies that have been developed, including cloud computing (Bairagi & Bang, 2015). The first concept of cloud computing was introduced in 2004, and it was defined by many academics and practitioners in commercials with different visions (Silverio et al., 2017). The word cloud in Cloud Computing means provider or carrier who provides service over the internet, while computing refers to the different resources that are provided by a computer (Surbiryala & Rong, 2019). Cloud Computing is a term for networked computers that deliver informational technology services over the internet to many users in an on-demand environment. These can be customised to the client's needs (Thomas, 2010). It is a model for enabling convenient, on-demand network access to a shared pool of configurable computing resources (e.g., networks, servers, storage, applications, and services) that can be rapidly provisioned and released with minimal management effort or service provider interaction (Bello et al., 2021). Cloud computing involves the provision of informational technology as a service, not as a product (Senyo et al., 2016).

Abedi et al. (2014) submitted that cloud computing is a virtual technology that sends and retrieves data and different applications by using the internet and central remote servers, including application and database servers. It is a promising technological advancement that will change the construction industry. According to Afolabi et al. (2017), cloud computing is not entirely new to the construction participants, as it has been optimised in the form of Microsoft Office Live, Dropbox and Google Drive. This technology offers the enhancement of schedule performance, stakeholder engagement, better quality, and reduced rework (Bademosi & Issa, 2021). Oke et al. (2021) affirm that when cloud computing is implemented in a construction project, the cost of data will be reduced, construction management will be enhanced, project planning will be improved, and project progress can be monitored. Moreover, cloud computing offers scalability and flexibility in the delivery of construction projects and, at the same time, offers an opportunity for improved security against data loss and manipulation (Bello et al., 2021).

Three-dimensional (3D) Printing

Three-dimensional (3D) printing, also known as additive manufacturing, is an automated process that requires software, hardware, and materials to produce a solid final product. A variety of 3D software programs in the construction market can be used to create a 3D digital model. Once the digital model has been created, it is scanned with a 3D scanner. The 3D scanner processes the digital model into a readable file format, which the 3D printer can process. The 3D printer then extrudes printable material on site in accordance with the layers and processes of the 3D digital model (Lim et al., 2012; Sakin & Kiroglu, 2017).

Contour crafting is the technology deployed in the construction industry for 3D-printed houses. According to Wu et al. (2016), contour crafting can produce solid structures that are greater than 1 m and involves the process of extruding cement base paste from the 3D printer nozzle to form the internal and external skeleton walls of the house. The inside of the walls is filled with a bulk compound similar to concrete. Contour crafting uses two basic methods vis: Fuse Deposition

Modeling (FDM) and repeating layers until the whole product is complete. Literature sources credit Scott Crump for the invention of FDM technology in 1988. FDM requires ductile materials that self-harden during the cooling process to be extruded by the 3D printer's first nozzle and deposition of a supportive material by the second nozzle. The two nozzles are located in the printer head of the robotic arm. The robotic arm of the 3D printer deposits the ductile materials from its nozzles by following the cartesian coordinates set out in the 3D digital model, hence the creation of cross-sectional layers (Hager et al., 2016).

Since its first invention in the 1980s, the developments in 3D printing have continued to gain prominence in diverse fields, and its advancements are growing rapidly (Hager et al., 2016). The construction industry has been noted to have stayed the same as 100 years ago regarding building techniques and materials. Developments in the commercialisation of 3D printing in the construction industry are becoming popular, thus increasing the potential to change the construction industry landscape (Sakin & Kiroglu, 2017). The number of new entrants into the 3D printing market is increasing. In 2013, it was recorded that there were only approximately 20 start-ups in the 3D printing construction field. However, most recent reports list about 65 business entities in 3D printing in the construction industry that offer services ranging from prototyping solutions, software and design tools, 3D printing of large components or 3D printing of entire buildings (Laubie et al., 2018). The benefits of using 3D printing of houses include increased durability and quality of houses through better selection of material based on life cycle and the automated process of workmanship (Sakin & Kiroglu, 2017). The construction costs are significantly reduced because the materials required to build the framework of the building are printed on-site, and costs are lower than traditional methods. There is a reduction in material transportation and storage needs. In addition, there is a reduction in construction waste material being produced, and printers can do the most hazardous and dangerous construction work, thus lowering injuries on site. The time required to produce a single building will be reduced (Hager et al., 2016).

The immense benefits inherent in using 3D printing in the construction industry have led to the widespread use of this technology. Goldin (2014) noted that 3D printing technology was utilised in China to construct ten one-storey houses within 24 hours. A similar observation was noted in Dubai by MacRae (2016), where the 'office of the future' was constructed within 19 days, out of which 17 days were used for printing the building components and 2 days to assemble the structure on-site. The building was initially used as a meeting space. The 3D-printed office is a fully operating structure with services like electricity, water, telecommunications, and air-conditioning systems. In France, this technology was adopted to deliver a 1000 $ft.^2$ home as a model for social housing. The delivery of this project took about four and a half days to complete. It has been noted that the government in Dubai envision the use of 3D printing to solve its problem of housing shortage by the year 2025 (Jones Lang LaSalle, 2021). Icon in the USA also developed its first industrial 3D printer called the 'Valcan II', designed to print the first low-income houses in El Salvador. In Chennai, India, a $56m^2$ home was constructed within 5 days, while in the Netherlands, a five-house project was completed with occupants already using the structure (Fleming, 2021). 3D printing promises rapid prototyping and mass customisation of construction works (Wu et al., 2016), reduction in the cost of labour and material needs for delivering construction projects (Hager et al., 2016; Lim et al., 2012; Mohd-Tobi et al., 2018; Sakin & Kiroglu, 2017), reduction in transportation costs and required storage areas, reduction in site injuries and fatalities (Hager et al., 2016; Sakin & Kiroglu, 2017), better design creativity (Labonnote et al., 2016), creation of new industries and new jobs (Fonseca, 2018) and improved productivity and performance in the construction industry (Aghimien et al., 2020).

Digital Twin

A digital twin is a live, virtual model that accurately reflects a physical asset or system and can adapt to changes in the environment or operations (Gopinath et al., 2019). It allows users to illustrate how a product would perform before construction, as well as analyse the past and present, and forecast the future state of physical objects (Johnson Controls, 2019). The ability to forecast future states and processes using digital twins helps reduce risks and save costs (BSI, 2018). Digital twin is an extension of BIM, created using a combination of existing technologies and initiatives (BSI, 2018). In the construction industry, the digital twin has several applications, including being able to predict faults in equipment, provide real-time operational views, improve cost estimates and reduction, mitigate risks and enhance security (Johnson Controls, 2019).

Blockchain Technology

Blockchain is a distributed, shared digital ledger that records information in a permanent and immutable way across multiple sites and entities (Hultgren & Pajala, 2018; Penzes, 2018). Blockchain technology plays a significant role in enhancing the procurement process. It allows transparency and easy access to track materials and work activities in the supply chain (Sahu, 2023). Moreover, efficiency and speed in the procurement process can be attained through this technology by streamlining the entire process. At the construction stage, blockchain technology can help achieve effective accountability on the part of all parties involved in the project. It also affords transparency and seamless collaboration between project stakeholders (Panchal, 2022). Moreover, potential risks can be avoided as the system allows real-time monitoring and early identification of possible risks that could have affected the project. In terms of operation, blockchain technology offers efficient management of projects through real-time tracking and lifecycle management. Moreover, the system affords optimum security of sensitive operational data and transactions, thereby protecting them from unauthorised access (Sahu, 2023). Overall, it is useful in monitoring the performance of the project to ensure effective maintenance management (Ngcobo et al., 2024).

While blockchain technology is being adopted in the construction industry, its implementation is still in the early stages and faces multiple challenges (Penzes, 2018). Stakeholders need to collaborate in new ways to address industry-wide issues. Potential applications of blockchain in construction include contract management, electronic document management, property management, supply chain management, and funding management (Penzes, 2018; San et al., 2019). Blockchain can increase transparency by verifying the origin and handling of materials, which could help curb corruption and falsified certifications (Hultgren & Pajala, 2018). Key areas of application include procurement, supply chain, payments, project management, BIM, and smart asset management. Overall, blockchain offers promising solutions for the construction industry, but widespread adoption is still in the early stages and requires collaboration among stakeholders.

Digitalisation in the Construction Industry

Developed Countries Perspective

Although the use of digital technologies in the construction industry of developed countries is still slow-paced when compared to other industries like manufacturing, agriculture, finance and education (Aghimien et al., 2021), a significant level of adoption and support for the use of emerging technologies have been noted. For instance, in the USA, the emergence of digital technologies such as cloud computing has received significant support from the General Services Administration through the Federal Cloud Computing Initiative (Kumar & Cheng, 2010). In the UK, the government is at the forefront of promoting emerging digital technologies like BIM cloud computing, among others (Lumsden & Gutierrez, 2013). BIM has found its way into leading construction and engineering organisations in countries like Australia, Finland, North America, and the UK (Aghimien et al., 2021). Haron et al. (2010) observed that most organisations within the built environments around the world, particularly in developed countries, are moving towards BIM adoption in their practices. As such, researchers and establishments have aspired to quantify the status of BIM appropriation since this is basic in assessing BIM adoption and other issues surrounding technological innovation. This understanding has prompted reports such as the SmartMarket arrangement, National Building Specifications (NBS) reports, BIM overviews, and others. These reports mostly studied respondents' involvement with BIM, the advantages and future significance of BIM, possible utilisation of BIM programming applications, and arranged BIM speculation. Based on these overviews, the selection rate and experience level of respondents in North America quickly expanded from 2009 to 2012, and the respondents from South Korea and Oceania anticipated that their profundity of BIM execution would increase sooner rather than later (Jung & Lee, 2015). The UK's BIM Report 2020 noted a rise in the use of the technology, with over 73% of sampled organisations stating they now use BIM in their project delivery in the country. The HM Government further noted that BIM is a crucial part if the UK's construction strategy, and this has put the country at the forefront of global BIM standards and adoption (HM Government, n.d.; Trimble, 2023).

Similarly, in Germany, the government has been proactively pushing the adoption of digital technologies. In fact, the government mandated the use of BIM for all public projects by 2020 (Soto, 2024). In Japan, the government expected a 50% adoption rate of BIM by 2020 (Pinnacle, 2023). Moreover, Ragpala (2024) observed that the continuous demand for construction products in Japan has become a critical driver for the high demand for cloud-based engineering technologies currently being experienced in the country. In Denmark, the government has continued to invest in emerging digital technologies as they recognise that they are germane to the country's digital infrastructure for both the public and private sectors (Ministry of Foreign Affairs of Denmark, 2024). In Australia, Ai Group (2024) submitted that there has been a significant increase in the use of digital technologies, with about 84% of business leaders indicating their use of diverse emerging technologies. Similarly, in 2023, Autodesk commissioned Deloitte Access Economics to investigate the impact of digital technologies on the construction and engineering sector. It was observed that many construction businesses have had to restructure their business plans in response to technological advancements, which has led to increased use of technologies. Evidence of the use of these technologies has been noted in dispute resolution (Golem et al., 2024). Draft Tech (2024) gave a snapshot of digital technology adoption in the Australian construction industry. It was noted that digital twins, data analytics, and AI have all become prominent in addressing the industry's issues, particularly for remote working, tackling labour shortage, and driving meaningful change within the industry.

It is evident that the government in many developed countries are at the forefront of driving the digital technology uptake in their construction industry through investments, standards and legislation. This is expected as the government is the major client of the construction industry globally, and they need the industry to actualise their role and responsibility of providing quality infrastructure for their citizens (Aghimien et al., 2021).

Developing Countries Perspective

The construction industry in developing countries has been slow in the uptake of emerging digital technologies, and this has immensely influenced the slow-paced digitalisation process, which is evident in many of these countries (Aghimien et al., 2021). Studies have noted that developing countries face unique problems with the implementation of digital technologies due to several factors, including a lack of resources, inadequate supporting infrastructure and diverse levels of readiness to adopt technologies (Rinchen et al., 2024; Stojanovska-Georgievska et al., 2022; Takyi-Annan & Hong, 2023). However, Othman (2023) noted that many developing countries have the opportunity to 'leapfrog' into the digital era by employing collaborative approaches, fostering a culture of innovation, ensuring digital skills are developed, reforming existing policies and regulations, ensuring access to finance, addressing infrastructure deficits, and showcasing case studies and success stories.

In India, there is a widespread awareness of emerging technologies and digitalisation across diverse construction organisations. However, implementing these technologies is dismal (Dauda et al., 2024). In Malaysia, Gamil et al. (2020) explored the use of digital tools like IoT in the construction industry and observed despite its inherent benefits and awareness among practitioners, its adoption is still limited to a few applications. The situation is the same in many other developing countries (Ikuabe et al., 2021; Oke et al., 2023), where awareness of digitalisation is rapidly increasing among construction professionals, but the actual implementation is very poor. Hence, the slow adoption of digital technologies in many developing countries is not an issue of lack of awareness but the proper approach to implementation, among other issues.

In South Africa, Osunsanmi et al. (2018) have earlier noted that there is an increase in awareness among South African construction professionals on the fourth industrial revolution principles. Many are willing to adopt digital technologies and concepts of this industrial revolution in their daily construction processes. This is because these digital technologies offer developing sustainable buildings, efficient communication among project teams and better project time and cost efficiency. Similarly, Aghimien et al. (2021) found a certain level of usage of digital technologies in the South African architecture, engineering and construction industry. Most digital technologies are dominantly used during the design and feasibility stages for cost control and monitoring. In Nigeria, Oke et al. (2023) explored the application of digital technologies, particularly in social and sustainable construction. The study noted that designing, cost planning, analysis of building systems and structural elements, and contract documentation are areas where emerging digital tools are mostly used in the country's construction industry. From the perspective of information management in construction, Adekunle et al. (2024) submitted that experts in the South African construction space noted that digital technology adoption offers better communication, operational efficiency, and market intelligence.

In Vietnam, Van Tam et al. (2024) noted that while digital technologies offer the opportunity to improve the efficiency and quality of construction, their adoption is still slow-paced due to unique issues facing the country's construction industry. It was further noted that social and habitual resistance to change was the leading factor of the slow adoption of merging digital tools in the country. This problem counters the effort of construction organisations in the country who are seeking ways to optimise their operations and leverage emerging digital tools to mitigate risks, streamline processes, and maximise overall project outcomes (Adomako & Nguyen, 2023; Cuong et al., 2023; Hai et al., 2022). In the Dominican Republic, Reyes-Veras et al. (2021) explored the application of emerging technologies like big data analytics and observed sow integration of the technology into construction activities. It was further noted that this slow integration can be attributed to awareness issues, high cost of investment, resistance to change, lack of government support and regulation, lack of technological expertise, and security concerns. However, effectively adopting this emerging technology can help the country's construction industry meet the demanding nature of construction project delivery.

Rinchen et al. (2024) explored the implementation of BIM in developing countries. It was noted that emerging technologies like BIM, for instance, are gradually gaining recognition within the construction industry of many developing countries, and the level of implementation is still low. BIM implementation level one has been noted in many countries like Nigeria, India, Pakistan, Algeria, and Egypt, among others (Mohanta & Das, 2022; Othman et al., 2021; Rinchen et al., 2024) with others like Vietnam, Malaysia, China, Brazil reaching highest of level two (Munianday et al., 2023; Rinchen et al., 2024). Further noted that some countries are even yet to implement, with a level zero being noted (Rinchen et al., 2024; Zaia et al., 2023). This is the case for other emerging technologies, many of which are still in their infancy in developing countries. As such, there is a need to understand the hindrances leading to this slow adoption and propose measures for improving digital technology uptake in the global construction industry.

Challenges and Measures for Improving Digitalisation in Construction

The application of digital technologies to attain the digitalisation of construction processes and services is challenged by several issues which vary from country to country. Gamil et al. (2020) noticed that the use of digital technologies in construction is challenged by safety and security issues, poorly documented standards, a lack of awareness of the benefits, the improper introduction of digital technologies, and a lack of robustness in connectivity. Dauda et al. (2024) found that issues relating to finance, organisational culture, regional disparities, data security, and capacity-building. It was further observed that larger organisations are more proactive in pursuing digital transformation, but smaller organisations struggle due to lacking resources and knowledge. An earlier study by Parwal (2023) noted that reluctance among stakeholders and lack of digital competence are other issues facing the digitalisation of the construction industry.

Struková and Líška (2012) explored the application of digital tools, such as robotics, in the construction industry and found that the high costs of acquiring and maintaining these technologies and the fact that they are unavailable locally are the two leading factors that make their implementation difficult. It was further noted that some of these emerging technologies are incompatible with the country's existing construction practices. A similar observation was made by Vaduva-Sahhanoglu et al. (2016), who noted that high initial costs, the fragmented nature of the construction market, high cost for research and development innovation, high cost for updating current technologies, costs for training employees in using the technologies and costs involved in changing construction operations are issues facing digital technology adoption in many countries around the world. The issues of high cost of acquiring and maintaining digital technologies have been noted across diverse countries in relation to diverse emerging technologies (Chimhundu, 2016; Golizadeh et al., 2019; Yahya et al., 2019).

In addition, issues regarding the difficulty in using and understanding the technologies have also been noted (Struková & Líška, 2012). According to Yahya et al. (2019), many digital technologies are more advanced and can be challenging to low- or semi-skilled workers in the construction industry. This complexity can lead to resistance to usage, especially in the much older workforce. Vaduva-Sahhanoglu et al. (2016) noted that resistance to accepting new technologies can be a major problem. Yahya et al. (2019) further pointed out that in some countries, worker unions view these technologies as replacements for workers and, as such, resist their implementation. Aghimien et al. (2024) also noted that fear of job loss has led to resistance to adopting digital technology in many developing countries.

Reyes et al. (2018) noted that emerging technologies like BDA are not being adopted in the construction industry of many countries due to a lack of interest and knowledge of its application. Knowledge and training issues were also highlighted by Chimhundu (2016) when exploring barriers to BIM adoption. It is important to have the right skills through effective training to implement emerging digital technologies successfully. Oke et al. (2018) noted the absence of specific modules designed towards digital training in many institutions. The implication is that construction organisations have to train their employees on the use of digital tools, thereby leading to more expenses and a critical challenge towards implementing these technologies.

Studies have also noted other issues like technical difficulties, including data loss, inaccuracy in data generated, and lack of internet connection, among others, can be critical to the successful implementation of digital technologies like UAVs, IoT and BDA (Bilal et al., 2016; Golizadeh et al., 2019). This situation is further complicated by security issues and fear of cyber-attacks (Aghimien et al., 2021). PricewaterhouseCoopers (2016) noted that in many cases, drones are used to collect data, which could sometimes include confidential, private, or sensitive information relating to private property or private behaviour. This security of private and sensitive information can be a problem when adopting the technology. The situation is no different with the use of IoT. Rad and Ahmad (2017) noted that end users tend to worry about the privacy of their devices and data as the IoT enables intercommunication between connected devices and humans. More so, personal, home and work devices being extensively exposed to the huge network of interconnected smart devices can create worry for users. From the perspective of BIM usage, Ghaffarianhoseini et al. (2016) mentioned that cyber security can be a problem that is not carefully addressed as BIM involves information sharing, which allows the project team to access project data. This can cause issues of copyright infringement and unauthorized access. Atobishi et al. (2018) further emphasised that security concerns such as phishing and data loss can be problematic when integrating cloud computing into construction. Other studies have also noted cyber security, privacy, protection, and data quality issues related to using BDA and cloud computing in construction (Bilal et al., 2016; Reyes et al., 2018). Other issues noted include restrictive and unsupportive regulations and legislations (Golizadeh et al., 2019), legal and contractual issues (Azhar, 2011; Sardroud et al., 2018), lack of awareness and demand (Kiprotich, 2014; Sawhney & Singhal, 2013) lack of organisational support (Holt, 2015; Kiprotich, 2014), interoperability and incompatibilities of systems (Ghaffarianhoseini et al., 2016; Li et al., 2018; Ratajczak et al., 2015).

Based on the aforementioned, the challenges facing the use of digitalisation in construction can be summarised into the following:

1. Absence of digital technology teaching in higher education.
2. Compatibility with existing construction practices.
3. Complexity of digital technologies.
4. Costs for training employees in using the technologies.
5. Data loss.
6. High cost for research and development innovation.
7. High cost for updating current technologies.
8. High costs of acquiring and maintaining digital technologies.
9. Improper introduction of digital technologies.
10. Inaccuracy in data generated.
11. Interoperability issues.
12. Lack of awareness and demand.
13. Lack of awareness of the benefits.
14. Lack of digital competence.
15. Lack of internet connection.
16. Lack of knowledge.
17. Lack of organisational support.
18. Lack of robustness in connectivity.
19. Lack of training.
20. Legal and contractual issues.
21. Poorly documented standards.
22. Reluctance among stakeholders.
23. Resistance to change.
24. Restrictive and unsupportive regulations and legislation.
25. Safety, security and fear of cyber-attack.
26. Unavailable locally.

Evidently, digitalisation of the construction process offers greater efficiency in construction delivery while ensuring projects are delivered on time, within budget and to the satisfaction of clients through effective communication between project participants (Aghimien et al., 2021). Thus, it is crucial to ensure construction digitalisation is achieved in the global construction industry by implementing measures to address the challenges facing digital technology deployment above. Existing literature suggests several measures for improving the use of digital technologies in construction. Babatunde et al. (2018) noted that support from the government through policies and legislation can prove highly influential in driving digital technology use in construction. Oke et al. (2018) further noted that through soft loans by the government for technological development and incentivising the adoption of digital technologies, the adoption level in construction can increase. Investment in training programmes to ensure that employees are knowledgeable and skilled in using digital technologies. This can help increase efficiency and productivity, which can lead to improved client satisfaction. Organizations' provision of essential skills to workers is crucial in enabling them to be empowered (Orando & Isabirye, 2018). Research has proven that staff training on digital technologies is significant in implementing digital technologies in the construction industry (Oke et al., 2018). Training or building staff involves improving the performance and knowledge of the employees of the overall company. One of the primary goals of any construction organisation is growth, and for this growth to occur, the performance of the workforce must be great. Therefore, training the workers on how to operate and better understand digital technologies is significant for construction firms (Ngwenya & Aigbavboa, 2017).

Also, there is a need for collaboration with stakeholders and other organisations to effectively use digital technologies. Aghimien et al. (2021) described this collaboration as digital partnering, which entailed combining two or more organisations to share resources to attain digitalisation. Wallen and Sodergren (2021) have also noted the importance of collaboration between construction organisations and various stakeholders. It was noted that this collaboration is a crucial necessity for achieving the desired project objectives through the pooling of resources. This highlights the importance of collaboration as a key strategy for enhancing the integration of digital technologies in construction. Conducting internal communication campaigns that reinforce employees' knowledge of digital tools or providing incentives to encourage their interest and dedication towards digital innovation can also help in promoting the use of emerging digital technologies. Pascucci et al. (2023) noted that it is essential for senior and top-level management to take a leading role in this initiative. This will allow organisations to ensure that their personnel are equipped with the necessary skills and knowledge to effectively use digital technologies, ultimately leading to improved client satisfaction. Also, studies have noted the need for forward-looking research and development (R&D). Meno (2020) noted that construction organisations and professional and regulatory bodies should make a deliberate effort to release more information from research on the available digital technologies to educate and create awareness among industry players, encouraging its adoption. This will also increase the funding opportunities for further research and development in digital technologies, which will enhance the industry's technological capabilities (Calitz & Wium, 2022).

Organisations must prioritise cybersecurity for the safe and secure use of digital technologies. This is because construction projects involve exchanging sensitive data such as project plans, financial information, intellectual property, and personally identifiable information. Ensuring the security of this data is essential to protect the interests of all project participants (Pärn & Edwards, 2019). Cyberattacks are on the rise, and the construction industry is not immune to this threat. Hackers may attempt to gain unauthorized access to construction systems, leading to potential disruptions, financial losses, and reputational damage. By prioritising data security, the industry can minimise the risk of cyberattacks and protect against potential breaches. Aghimien et al. (2021) proposed a capability maturity model for the digitalisation of construction organisations. The study theorised that a construction organisation that will attain digitalisation, provide better service delivery, and, in the process, obtain a competitive edge must consider its maturity in six dynamic digital capabilities, namely, technology, people, process, strategy, digital partnering, and environment. Through these dynamic digitalisation capabilities, managers within these organisations can effectively utilise both internal and external digital competencies.

Based on the review of existing studies, the measures for improving the digital technology uptake can be summarised as follows:

1. Collaboration with stakeholders.
2. Continuous training and development for workers.
3. Harmonisation of different construction professionals' practices.
4. Improving digital infrastructure.
5. Incentivise adoption.
6. Internal campaign to promote digital technologies among workers.
7. Investment in cyber security.
8. Investment in forward-looking R&D.
9. Promoting digital technology education.
10. Request from construction clients.
11. Senior-management support.
12. Standardisation of construction elements and processes.
13. Support from the government through policies, legislation and incentives.

Summary

This chapter explored digitalisation in the construction industry. The chapter concludes that the construction industry globally is stricken with several challenges, including the slow adoption of emerging digital technologies that have affected the industry's performance. When compared to other industries, the use of emerging digital technologies in the global construction industry is still slow. It was further noted that while the industry in developed countries like the UK, USA, Germany, Australia, Denmark and Japan spearheads the implementation of emerging digital technologies, the story is not the same for developing countries. The industry in most developing countries is still lagging in deploying emerging digital tools. For many countries, emerging technologies, such as BIM, 3D printing, and IoT, among others, are still in the infancy stages. However, the construction industry in developing countries stands a chance to leapfrog into the digital era by exploiting their inherent potential to employ collaborative approaches, foster a culture of innovation, ensure digital skills are developed, and reform existing policies and regulations. Some critical challenges affecting the deployment of digital technologies in construction include the cost of acquiring, using and maintaining digital technologies, security and privacy, interoperability and compatibility, among others. To address these challenges, several measures were proposed.

References

Abedi, M., Rawai, N. M., Fathi, M. S., & Mirasa, A. K. (2014). Cloud computing as a construction collaboration tool for precast supply chain management. *Jurnal Teknologi, 70*(7), 1–7.

Addo-Tenkorang, R., & Helo, P. T. (2016). Big data applications in operations/supply-chain management: A literature review. *Computers & Industrial Engineering, 101*, 528–543.

Adekunle, P., Aigbavboa, C., Akinradewo, O., Ikuabe, M., & Otasowie, K. (2024). Towards the uptake of digital technologies for construction information management: A partial least squares structural equation modelling approach. *Buildings, 14*, 827.

Adomako, S., & Nguyen, N. P. J. T. J. O. T. T. (2023). *Digitalization, inter-organizational collaboration, and technology transfer* (pp. 1–27). Springer.

Afolabi, A. O., Ojelabi, R. A., Fagbenle, O. I., & Mosaku, T. O. (2017). The economics of cloud-based computing technologies in construction project delivery. *International Journal of Civil Engineering and Technology, 8*(12), 233–242.

Aghimien, D., Aigbavboa, C., Aghimien, L., Thwala, W. D., & Ndlovu, L. (2020). Making a case for 3D printing for housing delivery in South Africa. *International Journal of Housing Markets and Analysis, 13*(4), 565–581. https://doi.org/10.1108/IJHMA-11-2019-0111

Aghimien, D. O., Aigbavboa, C. O., Oke, A. E., & Thwala, W. D. (2021). *Construction digitalisation—A capability maturity model for construction organisations*. Routledge.

Aghimien, L. M., Aigbavboa, C. O., & Aghimien, D. O. (2024). *Construction workforce management in the fourth industrial revolution era*. Emerald Publishing Limited.

Agyekum, K., Kukah, A. S., & Amudjie, J. (2022). The impact of COVID-19 on the construction industry in Ghana: The case of some selected firms. *Journal of Engineering, Design and Technology, 20*(1), 222–244.

Ahadzie, D. K. (2019). *Ghana's construction industry is lively but needs regulation. The conversation*. Retrieved August 20, 2020, from https://theconversation.com/ghanas-construction-industry-is-lively-but-needs-regulation-124733#:~:text=In%202016%20and%202017%2C%20construction,3%25%20of%20the%20labour%20force

Ai Group. (2024). *Technology adoption in Australian industry - Commercial, workforce and regulatory drivers*. Retrieved August 12, 2024, from https://www.aigroup.com.au/resourcecentre/research-economics/technology-adoption-in-australian-industry/#:~:text=Australian%20industry%20is%20very%20actively,businesses%20also%20reporting%20high%20rates

Aigbavboa, C. O., Aghimien, D. O., Thwala, W. D., & Ngozwana, N. (2022). Unprepared industry meets pandemic: COVID-19 and the South Africa construction industry. *Journal of Engineering Design and Technology, 20*(1), 183–200.

Alladi, T., Chamola, V., Sahu, N., & Guizani, M. (2020). Applications of blockchain in unmanned aerial vehicles: A review. *Vehicular Communications, 23*, 1–24. https://doi.org/10.1016/j.vehcom.2020.100249

Alshalalda, M. (2019). Causes of human population growth: A quantitative study. *Methodist University's Journal of Undergraduate Research and Creativity, 6*, 80.

Alsharef, A., Ovid, A., Uddin, S. M. J., & Albert, A. (2024). Biggest challenges facing the construction industry. In: *Proceedings of the Construction Research Congress, Des Moines, Iowa, March 20–23* (pp. 652–660).

Amri, T., & Marey-Pérez, M. (2020). Impact of COVID-19 on Oman's construction industry. *Technium Social Sciences Journal, 9*, 661–670.

Arayici, Y., & Aouad, G. (2010). Building information modelling (BIM) for construction lifecycle management. In *Construction and building: Design, materials, and techniques* (pp. 99–118). Nova Science Publishers.

Arndt, C., Davies, R., Gabriel, S., Harris, L., Makrelov, K., Modise, B., Robinson, S., Simbanegavi, W., van Seventer, D., & Anderson, L. (2020). *Impact of COVID-19 on the South African economy*. Retrieved August 20, 2024, from https://sa-tied.wider.unu.edu/sites/default/files/pdf/SA-TIED-WP-111.pdf

Associated General Contractors of America. (2022). *The economic impact of construction in the United States and California*. Retrieved August 12, 2024, from https://www.adc.org

Atobishi, T., Gábor, S. Z., & Podruzsik, S. (2018). Cloud computing and big data in the context of industry 4.0: Opportunities and challenges. In: *Proceedings of the international academic conferences*, Sevilla, 5–8 March 2018 (pp. 9–15).

Atzori, L., Iera, A., & Morabito, G. (2010). The internet of things: A survey. *Computer Networks, 54*(15), 2787–2805.

Australian Government. (2024). *Labour market insights: Construction industry*. Retrieved August 12, 2024, from https://www.jobsandskills.gov.au/data/labour-market-insights/industries/construction#:~:text=Construction%20is%20a%20large%20employing,demolition%2C%20clearing%20and%20site%20preparation

Ayodele, E. O., & Alabi, O. M. (2011). Abandonment of construction projects in Nigeria: Causes and effects. *Journal of Emerging Trends in Economics and Management Sciences, 2*(2), 142–145.

Azhar, S. (2011). Building information modeling (BIM): Trends, benefits, risks, and challenges for the AEC industry. *Leadership and Management in Engineering, 11*(3), 241–252.

Babatunde, S. O., Ekundayo, D., Babalola, O., & Jimoh, J. A. (2018). Analysis of the drivers and benefits of BIM incorporation into quantity surveying profession: Academia and students' perspectives. *Journal of Engineering, Design and Technology, 16*(5), 750–766.

Bademosi, F., & Issa, R. R. A. (2021). Factors influencing adoption and integration of construction robotics and automation technology in the US. *Journal of Construction Engineering and Management, 147*(8), 1–13.

Bahrin, M. A. K., Othman, M. F., Nor, N. H., & Azli, M. F. T. (2016). Industry 4.0: A review on industrial automation and robotics. *Jurnal Teknologi (Sciences and Engineering), 78*, 137–143.

Bairagi, S. I., & Bang, A. O. (2015). Cloud computing: History, architecture, security issues. *International Journal of Advent Research in Computer and Electronics, 28*, 102–108.

Barton, B. (2023). *The 6 key issues in the construction industry*. Retrieved December 2, 2023, from https://www.linkedin.com/pulse/6-key-issues-construction-industry-bill-barton/

Bataw, A., Kirkham, R., & Lou, E. (2016). The issues and considerations associated with BIM integration. *MATEC Web of Conferences, 66*, 8.

Bello, S. A., Oyedele, L. O., Akinade, O. O., Bila, M., Delgado, J. M. D., Akanbi, L. A., Ajayi, A. O., & Owolabi, H. A. (2021). Cloud computing in the construction industry: Use cases, benefits and challenges. *Automation in Construction, 122*, 1–18.

Bikitsha, L., Ngomane, N., & Mamafha, K. (2014). *Understanding the use of emotional intelligence during the project leadership process: A case of PMs*.

Bilal, M., Oyedele, L. O., Qadir, J., Munir, K., Ajayi, S. O., Akinade, O. O., Owolabi, H. A., Alaka, H. A., & Pasha, M. (2016). Big data in the construction industry: A review of present status, opportunities, and future trends. *Advanced Engineering Informatics, 30*(3), 500–521.

BIM Engineering. (2018). *Top 8 applications of IoT in construction industry*. Retrieved August 12, 2024, from https://medium.com/@bimengus2017/top-8-applications-of-iot-in-construction-industry-d08dc3fbe2a6

Blackridge. (2024). *Major challenges facing the construction industry and some possible solutions*. Retrieved August 12, 2024, from https://www.blackridgeresearch.com/blog/major-challenges-in-construction-industry-business

Bock, T. (2008). Construction automation and robotics. In C. Balaguer & M. Abderrahim (Eds.), *Robotics and automation in construction* (pp. 21–42). IntechOpen. Retrieved August 12, 2024, from https://api.intechopen.com/chapter/pdf-preview/5556

Bohari, A. A. M., Skitmore, M., Xia, B., Teo, M., Zhang, X., & Adham, K. N. (2015). The path towards greening the Malaysian construction industry. *Renewable and Sustainable Energy Reviews, 52*, 1742–1748.

BSI. (2018). *Digital twins for the built environment*. Retrieved October 19, 2019, from https://www.bsigroup.com/globalassets/localfiles/en-gb/events/standards-forum-november-2018/digital-twins.pdf

Buehler, M., & Gerbert, P. (2017). *How Elon Musk and other pioneers are shaking up the construction industry*. Retrieved March 10, 2023, from https://www.weforum.org/agenda/2017/03/elonmusk-innovation-construction-industry

Burger, R. (2017). *How the construction industry is using big data*. Retrieved August 12, 2024, from https://www.thebalance.com/how-the-construction-industry-is-using-big-data-845322

Buvik, M. P., & Rolfsen, M. (2015). Prior ties and trust development in project teams: A case study from the construction industry. *International Journal of Project Management, 33*(7), 1484–1494.

Cabrera-Sánchez, J., & Villarejo-Ramos, A. F. (2019). Factors affecting the adoption of big data analytics in companies. *Journal of Business Management, 596*, 415–429.

Cai, S., Ma, Z., Skibniewski, M., Guo, J., & Yun, L. (2018). Application of automation and robotics technology in high-rise building construction: An overview. In: *Proceedings of the 35th international symposium on automation and robotics in construction*, Berlin, Germany, 20–25 July.

Calitz, S., & Wium, J. A. (2022). A proposal to facilitate BIM implementation across the South African construction industry. *Journal of the South African Institution of Civil Engineering, 64*(4), 29–37.

Chahal, H., & Gulia, P. (2016). Big data analytics. *Research Journal of Computer and Information Technology Sciences, 42*, 1–4.

Chan, A. P., Javed, A. A., Lyu, S., Hon, C. K., & Wong, F. K. (2016). Strategies for improving safety and health of ethnic minority construction workers. *Journal of Construction Engineering and Management, 142*(9), 05016007.

Chan, D. W. M., & Aghimien, D. O. (2022). Bolstering measures for combating the challenges of safe working cycle implementation in Hong Kong's construction industry. *Sustainability, 14*(7), 1–18.

Chartered Institute of Building. (2018). *Socio-economic impact of construction*. Retrieved August 12, 2024, from https://www.ciob.org/industry/policy-research/policy-positions/socio-economic-impact-construction

Cheelo, C. & Liebenthal, R. (2018). The role of the construction sector in influencing natural resource use, structural change, and industrial development in Zambia. *WIDER working paper 2018/172*. Retrieved November 24, 2020, from www.wider.unu.edu/sites/default/files/Publications/Working-paper/PDF/wp2018-172.pdf

Chen, C. L. P., & Zhang, C. Y. (2014). Data-intensive applications, challenges, techniques and technologies: A survey on big data. *Information Sciences, 275*, 314–347.

Chen, X., Chang-Richards, A. Y., Pelosi, A., Jia, Y., Shen, X., Siddiqui, M. K., & Yang, N. (2022). Implementation of technologies in the construction industry: A systematic review. *Engineering, Construction and Architectural Management, 29*(8), 3181–3209.

Chi, H. L., Kang, S. C., & Wang, X. (2013). Research trends and opportunities of augmented reality applications in architecture, engineering, and construction. *Automation in Construction, 33*, 116–122.

Chimhundu, S. (2016). *A study on the BIM adoption readiness and possible mandatory initiatives for successful implementation in South Africa*. University of the Witwatersrand.

Cuong, N. M., Balaganesh, D., Bertero, L., & Mutai, N. J. J. (2023). Transformational leadership and digital transformation challenges: The case study of Deoca Group in Vietnam. *Journal of Social Sciences and Humanities Research, 35*, 1081–1105.

Datt, S. (2011). The information explosion: Trends in technology 2011 review. *The Journal of Government Financial Management, 60*(4), 46–51.

Dauda, J. A., Chavan, N. N., Saka, A. B., Ajayi, S. O., & Oyegoke, A. S. (2024). An appraisal of barriers to digitalisation of the construction industry in developing countries: Perspective from India. *International Journal of Construction Management*. https://doi.org/10.1080/15623599.2024.2362014

Davenport, T. H., & Patil, D. J. (2012). Data scientist: The sexiest job of the 21st century. *Harvard Business Review, 90*, 70–76.

Department of Statistics Malaysia. (2020). *Malaysia economic performance second quarter 2020*. Department of Statistics Malaysia official portal. Retrieved November 24, 2020, from www.dosm.gov.my/v1/index.php?r=column/cthemeByCat&cat=100&bul_id=NS9TNE9yeHJ1eHB6cHV1aXBNQlNUZz09&menu_id=TE5CRUZCblh4ZTZMODZIbmk2aWRRQT09

Dithebe, K., Aigbavboa, C., Oke, A., & Muyambu, M. A. (2018). Factors influencing the performance of the south African construction industry: A case of Limpopo Province. In *International conference on industrial Engineering and operations management*, Pretoria/Johannesburg, 29 October–1 November 2018 (pp. 1185–1192).

Dixit, S., Pandey, A. K., Mandal, S. T., & Bansal, S. (2017). A study of enabling factors affecting construction productivity: Indian scenario. *International Journal of Civil Engineering and Technology, 8*(6), 741–758.

Draft Tech. (2024). *The state of digital technology in the Australian construction industry: A 2024 snapshot by region*. Retrieved August 12, 2024, from https://www.draftech.com.au/the-state-of-digital-technology-in-the-australian-construction-industry-a-2024-snapshot-by-region/

Dubey, R. (2019). Big data analytics and organizational culture as complements to swift trust and collaborative performance in the humanitarian supply chain. *International Journal of Production Economics, 210*, 120–136.

Dupont, Q., Chua, D., Tashrif, A., & Abbott, E. (2017). Potential applications of UAV along the construction's value chain. *Procedia Engineering, 182*, 165–173. https://doi.org/10.1016/j.proeng.2017.03.155

Durdyev, S., & Ismail, S. (2012). Pareto analysis of on-site productivity constraints and improvement techniques in construction industry. *Scientific Research and Essays, 7*(4), 824–833.

Edirisinghe, R. (2018). Digital skin of the construction site: Smart sensor technologies towards the future smart construction site. *Engineering, Construction and Architectural Management, 26*(2), 184–223.

Elghaish, F., Matarneh, S., Talebi, S., Kagioglou, M., Hosseini, M. R., & Abrishami, S. (2021). Toward digitalisation in the construction industry with immersive and drone technologies: A critical literature review. *Smart and Sustainable Built Environment, 10*(3), 345–363.

European Commission. (2020). *Construction sector*. Retrieved August 12, 2024, from https://ec.europa.eu/growth/sectors/construction_en

Eventura. (2023). *Construction industry trends and challenges in 2023*. Retrieved August 12, 2024, from https://eventura.com/construction/construction-industry-trends-and-challenges-in-2023/

Fleming, S. (2021). *These 3 countries are 3D-printing new homes*. Retrieved August 12, 2024, from https://www.weforum.org/agenda/2021/05/countries-3d-printing-new-homes/

Fonseca, L. M. (2018). Industry 4.0 and the digital society: Concepts, dimensions and envisioned benefits. *Proceedings of the International Conference on Business Excellence, 34*, 386–397.

Gamil, Y., Abdullah, A., Abd Rahman, I., & Asad, M. M. (2020). Internet of things in construction industry revolution 4.0: Recent trends and challenges in the Malaysian context. *Journal of Engineering, Design and Technology, 18*(5), 1091–1102.

Gamil, Y., & Alhagar, A. (2020). The impact of pandemic crisis on the survival of construction industry: A case of COVID-19. *Mediterranean Journal of Social Sciences, 11*(4), 122–128.

Gartner. (2012). *Gartner IT glossary - Big data.* Retrieved August 12, 2024, from http://www.gartner.com/it-glossary/big-data

Ghaffarianhoseini, A., Tookey, J., Ghaffarianhoseini, A., Naismith, N., Azhar, S., Efimova, O., & Raahemifar, K. (2016). Building information modelling (BIM) uptake: Clear benefits, understanding its implementation, risks, and challenges. *Renewable and Sustainable Energy Reviews, 75*, 1046–1053.

Ghasri, M., & Maghrebi, M. (2021). Factors affecting unmanned aerial vehicles' safety: A post-occurrence exploratory data analysis of drones' accidents and incidents in Australia. *Safety Science, 139*, 1–11.

Goessens, S., Mueller, C., & Latteur, P. (2018). Feasibility study for drone-based masonry construction of real-scale structures. *Automation in Construction, 94*, 458–480.

Goldin, M. (2014). *Chinese company builds houses quickly with 3D printing*. Retrieved August 15, 2018, from http://mashable.com/2014/04/28/3d-printing-houses-china/

Golem, A., Minamikata, N. & Reynolds, J. (2024). *The digital transformation in construction—What you need to know.* Retrieved August 12, 2024, from https://www.herbertsmithfreehills.com/insights/reports/inside-arbitration-issue-18/the-digital-transformation-in-construction-what-you-need-to-know#:~:text=A%20recent%20Deloitte%20report%20has,day%2Dto%2Dday%20operations.&text=These%20include%20increased%20productivity%20(34,increased%20staff%20safety%20(33%25)

Golizadeh, H., Hosseini, M. R., Edwards, D. J., Abrishami, S., Taghavi, N., & Banihashemi, S. (2019). Barriers to adoption of RPAs on construction projects: A task–technology fit perspective. *Construction Innovation, 19*(2), 149–169.

Gopinath, V., Srija, A., & Sravanthi, C. N. (2019). Re-design of smart homes with digital twins. *Journal of Physics: Conference Series, 1228*(1), 1–9.

Gupta, M., & George, J. F. (2016). Toward the development of a big data analytics capability. *Information & Management, 53*(8), 1049–1064.

Hager, I., Golonka, A., & Putanowicz, R. (2016). 3D printing of buildings and building components as the future of sustainable construction? *Procedia Engineering, 151*, 292–299.

Hai, D. T., Toan, N. Q., & Van Tam, N. (2022). Critical success factors for implementing PPP infrastructure projects in developing countries: The case of Vietnam. *Innovative Infrastructure Solutions, 7*(1), 1–13.

Haron, A. T., Marshell-Ponting, A., & Aouad, G. (2010). Building information modelling: Literature review on model to determine the level of uptake by organisations. In *Proceedings of the 18th CIB world building congress 2010*, the Lowry, Salford Quays, 10–13 May (pp. 168–184).

Haupt, T. C., & Harinarain, N. (2016). The image of the construction industry and its employment attractiveness. *Acta Structilia, 23*(2), 79–108.

Heimgartner, J. (2020). *Long-term effects of COVID-19 in the construction industry.* Retrieved August 12, 2024, from www.engineering.com/BIM/ArticleID/20445/Long-Term-Effects-of-COVID-19-in-the-Construction-Industry.aspx

HM Government. (n.d.). *Building information modelling: Industrial strategy: government and industry in partnership.* Retrieved August 12, 2024, from https://assets.publishing.service.gov.uk/media/5a79b2eae5274a18ba50e280/12-1327-building-information-modelling.pdf

Holt, E. A. (2015). Emerging technology in the construction industry: Perceptions from construction industry professionals. In: *Proceedings of the 122nd ASEE Annual Conference and Exposition, Seattle, 14–17 June 2015* (pp. 1–10). Retrieved August 12, 2024, from https://www.semanticscholar.org/paper/Emerging-Technology-in-the-Construction-Industry%3A-Holt-Benham/2e80b12d3cce4a6e5f76f78ca9cb52719bef6c50

Hultgren, M., & Pajala, F. (2018). *Blockchain technology in the construction industry: Transparency and traceability in supply chain.* Royal Institute of Technology. Retrieved August 12, 2024, from http://www.diva-portal.org/smash/get/diva2:1229861/FULLTEXT01.pdf

Ibrahim, K., Simpeh, F., & Adebowale, O. J. (2023). Benefits and challenges of wearable safety devices in the construction sector. *Smart and Sustainable Built Environment*. https://doi.org/10.1108/SASBE-12-2022-0266

Ibrahim, R., Roy, M. H., Ahmed, Z. U., & Imtiaz, G. (2010). Analysing the dynamics of the global construction industry: Past, present and future. *Benchmarking: An International Journal, 17*(2), 232–252.

Ikuabe, M., Aghimien, D., Aigbavboa, C., Oke, A., & Thwala, W. (2021). Site accidents in the South African construction industry: Cleaning the Augean stables. In J. I. Kantola, S. Nazir, & V. Salminen (Eds.), *Advances in human factors, business management and leadership* (Vol. 267). Springer.

International Labor Organisation. (2001). *The construction industry in the twenty-first century: Its image, employment prospects and skill requirements*. ILO. Retrieved August 12, 2024, from www.ilo.org/public/english/standards/relm/gb/docs/gb283/pdf/tmcitr.pdf

International Trade Administration. (2024). *Building & construction*. Retrieved August 12, 2024, from https://www.trade.gov/country-commercial-guides/australia-building-construction#:~:text=Overview,economy%20(Master%20Builders%20Australia)

Ismail, S. A., Bandi, S., & Maaz, Z. N. (2018). An appraisal into the potential application of big data in the construction industry. *International Journal of Built Environment and Sustainability, 5*(2), 145–154.

Jeong, E., Seo, J., & Wacker, J. (2020). Literature review and technical survey on bridge inspection using unmanned aerial vehicles. *Journal of Performance of Construction Facilities, 34*(6), 1–15.

Johnson Controls. (2019). *Applying digital twins to the built environment*. Retrieved October 19, 2019, from https://www.johnsoncontrols.com/-/media/jci/insights/2019/bts/jci-661_dv_digital_twin_white_paper_020819_6p_f3.pdf

Jones Lang LaSalle. (2021). *3D printed buildings push construction boundaries*. Retrieved August 12, 2024, from https://www.jll.com.hk/en/trends-and-insights/investor/3d-printed-buildings-push-construction-boundaries

Jung, W., & Lee, G. (2015). The status of BIM adoption on six continents. *World Academy of Science, Engineering and Technology International Journal of Civil and Environmental Engineering, 9*(5), 512–516.

Kadiri, Z. O., Nden, T., Avre, G. K., Oladipo, T. O., Edom, A., Samuel, P. O., & Ananso, G. N. (2014). Causes and effects of accidents on construction sites: A case study of some selected construction firms in Abuja F.C.T Nigeria. *IOSR Journal of Mechanical and Civil Engineering, 11*(5), 66–72.

Khan, M. (2004). *Globalisation and its impact on the labor markets*. Economic Vista, Quarterly Magazine - International. Retrieved June 2, 2019, from http://www.ecovista.netfirms.com/trade/globalisation.htm

Kiprotich, C. J. K. (2014). *An investigation on building information modelling in project management: Challenges, strategy and prospects in the Gauteng construction industry, South Africa*. University of the Witwatersrand.

Kodeboyina, S. M., & Varghese, K. (2016). Low-cost augmented reality framework for construction applications. In: *Proceedings of the international symposium on automation and robotics in construction (ISARC), Alabama, 18–21 July 2016*.

Kumar, B., & Cheng, J. C. P. (2010). Cloud computing and its implications for construction IT. In W. Tizani (Ed.), *Computing in civil and building engineering, proceedings of the international conference* (p. 315). Nottingham University Press.

Kumar, S. A., Vealey, T., & Srivastava, H. (2016). Security in the internet of things: Challenges, solutions, and future directions. In *2016 49th HI international conference on system sciences (HICSS)* (pp. 5772–5781). IEEE.

Kwon, O. S., Park, C. S., & Lim, C. R. (2014). A defect management system for reinforced concrete work utilizing BIM, image-matching and augmented reality. *Automation in Construction, 46*, 74–81. https://doi.org/10.1016/j.autcon.2014.05.005

Labonnote, N., Rønnquist, A., Manum, B., & Rüther, P. (2016). Additive construction: State of the art, challenges and opportunities. *Automation in Construction, 72*(3), 347–366.

Laubie, R., Wunder, M., Witthoft, S., & Rothballer, C. (2018). Will 3D printing remodel the construction industry? *Boston Consulting Group*, 1–18.

Lee, I., & Lee, K. (2015). The internet of things: Applications, investments, and challenges for enterprises. *Business Horizons, 58*, 431–440.

Leeds, R. (2017). *Top 4 challenges facing the construction industry*. Retrieved August 12, 2024, from https://www.futureofconstruction.org/blog/top-4-challenges-facing-the-construction-industry/

Lewis, S. (2023). *5 challenges facing the construction industry*. Retrieved August 12, 2024, from https://www.ideagen.com/thought-leadership/blog/five-challenges-construction-industry

Lewis, T. M. (2007). Impact of globalization on the construction sector in developing countries. *Construction Management and Economics, 25*(1), 7–23.

Li, X., Yi, W., Chi, H. L., Wang, X., & Chan, A. P. (2018). A critical review of virtual and augmented reality (VR/AR) applications in construction safety. *Automation in Construction, 86*, 150–162.

Li, Y., & Liu, C. (2018). Applications of multirotor drone technologies in construction management. *International Journal of Construction Management, 19*(5), 1–12.

Lim, S., Buswell, R. A., Le, T. T., Austin, S. A., Gibb, A. G. F., & Thorpe, T. (2012). Developments in construction-scale additive manufacturing processes. *Automation in Construction, 21*, 262–268.

Liu, Y., Dai, H., Wang, Q., Shukla, M., & Imran, M. (2020). Unmanned aerial vehicle for internet of everything: Opportunities and challenges. *Computer Communications, 155*, 66–83.

Lumsden, J. R. I., & Gutierrez, A. (2013). Understanding the determinants of cloud computing adoption within the UK. In: European, Mediterranean & Middle East Conference on Information System (pp. 1–16).

MacRae, M. (2016). *The 3D printed office of the future*. Retrieved December 21, 2024, from www.asme.org/topics-resources/content/3d-printed-office-the-future

Mahmud, S. H., Assan, L., & Islam, R. (2018). Potentials of internet of things (IoT) in the Malaysian construction industry. *Annals of Emerging Technologies in Computing (AETiC), 2*(4), 44–45.

Manyika, J., Chui, M., Brown, B., Bughin, J., Dobbs, R., Roxburgh, C., & Byers, A. H. (2011). *Big data: The next frontier for innovation, competition, and productivity*. McKinsey Global Institute.

Martinez, J. G., Gheisari, M., & Alarcon, L. F. (2020). UAV integration in current construction safety planning and monitoring processes: Case study of a high-rise building construction project in Chile. *Journal of Management in Engineering, 36*(3), 1–15.

Massay, C. (2024). *What are the biggest challenges in the construction industry?—& how to tackle them*. Retrieved August 12, 2024, from https://www.theaccessgroup.com/en-gb/construction/resources/challenges-in-construction-industry/

McAfee, A., & Brynjolfsson, E. (2012). Big data: The management revolution. *Harvard Business Review, 9*(1), 61–68.

Meno, T. (2020). *An assessment of risk associated with digitalization in the south African construction industry*. University of Johannesburg, South Africa.

Ministry of Foreign Affairs of Denmark. (2024). *Tech in Denmark—EU report on the state of the digital decade 2024*. Retrieved August 12, 2024, from https://investindk.com/insights/tech-in-denmark-eu-report-on-the-state-of-the-digital-decade-2024

Mitchell, A. (2020). *CLC statement on payment and contracts*. Retrieved August 12, 2024, from https://www.constructionleadershipcouncil.co.uk/news/clc-statement-on-payment-and-contracts

Mohanta, A., & Das, S. (2022). Causal analysis of slow BIM adoption in eastern India with a special focus on the green building sector. *Journal of the Institution of Engineers: Series A, 103*(1), 319–337.

Mohd-Tobi, A. L., Omar, S. A., Yehia, Z., Al-Ojail, S., Hashimi, A., & Orhan, O. (2018). Cost viability of 3D printed house in UK. *IOP Conference Series: Materials Science and Engineering, 319*(1), 1–8.

Moud, H. I., Flood, I., Zhang, X., Abbasnejad, B., Rahgozar, P., & McIntyre, M. (2021). Quantitative assessment of proximity risks associated with unmanned aerial vehicles in construction. *Journal of Management in Engineering, 37*(1), 1–13.

Munianday, P., Rahman, R. A., & Esa, M. (2023). Case study on barriers to building information modelling implementation in Malaysia. *Journal of Facilities Management, 21*(4), 511–534.

My Job Quote. (2024). *The latest construction industry statistics*. Retrieved August 12, 2024, from https://www.myjobquote.co.uk/blog/the-latest-construction-industry-statistics#:~:text=The%20percentage%20of%20people%20working,8.6%25%20of%20the%20total%20workforce

National Bureau of Statistics. (2020). *Nigerian Gross Domestic Product Report (Q3)*. Retrieved August 12, 2024, from www.nigerianstat.gov.ng/

Navon, R. (2005). Automated project performance control of construction projects. *Automation in Construction, 14*, 467–476.

Newton, E. (2023). *5 unique applications of IoT in construction*. Retrieved August 12, 2024, from https://www.iotforall.com/5-unique-applications-of-iot-in-construction

Ngcobo, N., Aigbavboa, C. O., & Aghimien, D. O. (2024). *Maintenance engineering for buildings: Digital solutions and sustainable practices*. Woodhead Publishing.

Ngowi, A., & Lema, N. (2002). Impact of globalization on the sustainability of construction firms in developing countries. In *Proceedings of CIB W107 1st international conference: Creating a sustainable construction industry in developing countries*, South Africa, November. Retrieved June 2, 2019, from http://www.odsf.co.za/cdcproc/docs/3rd/ngowi_lema.pdf

Ngwenya, L. M., & Aigbavboa, C. (2017). Improvement of productivity and employee performance through an efficient human resource management practice. *Advanced Intelligent Systems, 498*, 727–737.

Nizam, T., & Hassan, S. I. (2017). Big data: A survey paper on big data innovation and its technology. *International Journal of Advanced Research in Computer Science, 8*(5), 2173–2177.

Nuzha, R. (2023). *Construction definition, types & categories*. Retrieved May 9, 2024, from https://study.com/academy/lesson/construction-definition-types.html#:~:text=Construction%20Definition-,Construction%20is%20the%20process%20where%20contractors%20build%20structures%20that%20serve,infrastructure%2C%20dams%2C%20and%20railways

Office of National Statistics. (2020). *Construction output in Great Britain: April 2020*. Retrieved August 12, 2024, from https://www.ons.gov.uk/businessindustryandtrade/constructionindustry/bulletins/constructionoutputingreatbritain/april2020

Ofori, G. (2000). Challenges of construction industries in developing countries: Lessons from various countries. In: *Proceedings of the 2nd international conference on construction in developing countries: Challenges facing the construction industry in developing countries, Botswana, November*. Retrieved June 2, 2019, from http://buildnet.csir.co.za/cdcproc/docs/2nd/ofori_g.pdf

Ofori, G. (2018). Construction in developing countries: Need for new concepts. *Journal of Construction in Developing Countries, 23*(2), 1–6.

Oke, A. E., Aigbavboa, C., Aghimien, D. O., & Koloko, N. (2018). Challenges of digital collaboration in the South African construction industry. In: *Proceedings of the international conference on industrial Engineering and operations management*, Bandung, 6–8 march 2018 (pp. 2472–2482).

Oke, A. E., Aliu, J., & Onajite, S. (2023). Barriers to the adoption of digital technologies for sustainable construction in a developing economy. *Architectural Engineering and Design Management, 20*(3), 431–447.

Oke, A. E., Kineber, A. F., Al-Bukhari, I., Famakin, I., & Kingsley, C. (2021). Exploring the benefits of cloud computing for sustainable construction in Nigeria. *Journal of Engineering, Design, and Technology, 21*(4), 973–990.

Oladinrin, T. O., Ogunsemi, D. R., & Aje, I. O. (2012). Role of construction sector in economic growth: Empirical evidence from Nigeria. *FUTY Journal of the Environment, 7*(1), 50–60.

Orando, M., & Isabirye, A. (2018). Construction workers' skill development: A strategy for improving capacity and productivity in South Africa. *International Journal of Economics and Finance Studies, 10*(1), 66–80.

Osunsanmi, T. O., Aigbavboa, C. O., & Oke, A. E. (2018). Construction 4.0: The future of the south African construction industry. *World Academy of Science, Engineering and Technology international Journal of Civil and Environmental Engineering, 12*(3), 206–212.

Othman, I., Al-Ashmori, Y. Y., Rahmawati, Y., Amran, Y. M., & Al-Bared, M. A. M. (2021). The level of building information modelling (BIM) implementation in Malaysia. *Ain Shams Engineering Journal, 12*(1), 455–463.

Othman, U. (2023). *Leapfrogging into the digital era: Can the African construction industry do it?* Retrieved August 12, 2024, from https://womeninbim.org/insights/leapfrogging-into-the-digital-era-can-the-african-construction-industry-do-it/

Panchal, K. (2022). *Blockchain in construction: How will it change the industry in 2022?* Retrieved August 12, 2024, from https://www.ifourtechnolab.com/blog/blockchain-in-construction-how-will-it-change-the-industry-in-2022#:~:text=It%20helps%20keep%20the%20supply%20chain%20accountable,to%20identify%20fraud%2C%20theft%2C%20or%20other%20issues

Pärn, A., & Edwards, D. (2019). Cyber threats confronting the digital built environment: Common data environment vulnerabilities and blockchain deterrence. *Engineering, Construction and Architectural Management, 26*(2), 245–266.

Parwal, U. (2023). *India construction industry: Current status & trend analysis*. Retrieved August 12, 2024, from https://equipmenttimes.in/india-construction-industry-current-status-trend-analysis/

Pascucci, F., Savelli, E., & Gistri, G. (2023). How digital technologies reshape marketing: Evidence from a qualitative investigation. *Italian Journal of Marketing*, 27–58.

Paul, C. A., Aghimien, D. O., Ibrahim, A. D., & Ibrahim, Y. M. (2021). Measures for curbing unethical practices among construction industry professionals: Quantity surveyors' perspective. *Construction Economics and Building, 21*(2), 1–17.

Peiffer, E. (2016). *3 concepts that will shape the future of construction.* Retrieved August 12, 2024, from https://www.constructiondive.com/news/3-concepts-that-will-shape-the-future-of-construction/417049/

Penzes, B. (2018). *Blockchain technology in the construction industry: Digital transformation for high productivity*. Institute of Civil Engineers. Retrieved August 12, 2024, from https://www.ice.org.uk/ICEDevelopmentWebPortal/media/Documents/News/Blog/Blockchain-technology-in-Construction-2018-12-17.pdf

Pinnacle. (2023). *BIM's role in Japanese smart building technology*. Retrieved August 12, 2024, from https://pinnacleinfotech.com/bims-role-in-japanese-smart-building-technology/#:~:text=One%20notable%20initiative%20in%20Japan's,BIM%20in%20public%20works%20projects

PricewaterhouseCoopers. (2016). *Clarity from above: PwC global report on the commercial applications of drone technology*. Retrieved August 17, 2019, from https://pwc.blogs.com/files/clarity-from-above-pwc.pdf

Propeller Aero. (2024). *Drones in construction—Why they are beneficial and how to use them.* Retrieved August 12, 2024, from https://www.propelleraero.com/blog/drones-in-construction-why-they-are-beneficial-and-how-to-use-them/

Rad, B. B., & Ahmad, H. A. (2017). Internet of things: Trends, opportunities, and challenges. *International Journal of Computer Science Network Security, 17*(7), 89–95. Retrieved from https://www.researchgate.net/publication/326919200_Internet_of_Things_Trends_Opportunities_and_Challenges

Ragpala, K. P. (2024). *Cloud-based engineering: Japan's response to massive construction expansion*. Retrieved August 12, 2024, from https://openit.com/cloud-based-engineering-japans-response-to-massive-construction-expansion/

Ratajczak, J., Malacarne, G., Krause, D., & Matt, D. (2015). The BIM approach and stakeholder's integration in the AEC sector: Benefits and obstacles in the south Tyrolean context. In: *Proceedings of the 4th international workshop on Design in Civil and Environmental Engineering*, Taiwan, 30–31 October 2015 (pp. 32–40).

Reyes, P., Suresh, S., & Renukappa, S. (2018). The adoption of big data concepts for sustainable practices implementation in the construction industry. In: *Proceedings of the 2018 IEEE/ACM international conference on utility and cloud computing companion*, Zurich, 17–20 December 2018. https://doi.org/10.1109/UCC-Companion.2018.00079

Reyes-Veras, P. F., Renukappa, S., & Suresh, S. (2021). Challenges faced by the adoption of big data in The Dominican Republic construction industry: An empirical study. *Journal of Information Technology in Construction, 26*, 812–831.

Rinchen, S., Banihashemi, S., & Alkilani, S. (2024). Driving digital transformation in construction: Strategic insights into building information modelling adoption in developing countries. *Project Leadership and Society, 5*, 1–20.

Rudyi, V. (2024). *Applying IoT in construction: Top 7 use cases in 2024 and beyond*. Retrieved August 12, 2024, from https://www.agilevision.io/blog/applying-iot-in-the-construction-industry-top-7-use-cases/

Rüßmann, M., Lorenz, M., Gerbert, P., & Waldner, M. (2015). *Industry 4.0: The future of productivity and growth in manufacturing industries*. Boston Consulting Group.

Ryfan. (2018). *Why construction matters*. Retrieved May 9, 2024, from https://www.ryfan.ca/company/construction-connundrum

Sahu, M. (2023). *5+ advantages of blockchain technology with real-life examples*. Retrieved August 12, 2024, from https://www.upgrad.com/blog/advantages-of-blockchain-technology/

Sakin, M., & Kiroglu, Y. C. (2017). 3D printing of buildings: Construction of the sustainable houses of the future by BIM. *Energy Procedia, 134*, 702–711.

San, K. M., Choy, C. F., & Fung, W. P. (2019). The potentials and impacts of blockchain technology in the construction industry: A literature review. In: *Proceedings of the 11th Curtin University technology, science and Engineering (CUTSE) international conference*, Sarawak, 26–28 November 2018, pp. 1–9.

Sardroud, J. M., Mehdizadeh Tavasani, M., Khorramabadi, A., & Ranjbardar, A. (2018). Barriers analysis to effective implementation of BIM in the construction industry. In: *35th international symposium on automation and robotics in construction (ISARC)*, Berlin, Germany, 20–25 July.

Sawhney, A., & Singhal, P. (2013). Drivers and barriers to the use of building information modelling in India. *International Journal of 3-D Information Modeling, 2*(3), 46–63.

Senyo, P. K., Affah, J., & Addae, E. (2016). Preliminary insight into cloud computing adoption in a developing country. *Journal of Enterprise Information Management, 29*(4), 505–524.

Shvets, Y., & Hanák, T. (2022). Use of the internet of things in the construction industry and facility management: Usage examples overview. *Procedia Computer Science, 219*, 1670–1677.

Silverio, M., Renukappa, S., Suresh, S., & Donastorg, A. (2017). Mobile computing in the construction industry: Main challenges and solution. In *Proceedings in business and economics* (pp. 85–99). Springer.

Skudder, A. (2021). *Construction industry a potential flywheel for economic recovery in SA*. Retrieved April 30, 2023, from https://www.engineeringnews.co.za/print-version/construction-industry-a-potential-flywheel-for-economic-recovery-in-sa-2021-11-18

Soto, C. (2024). *Construction technology: 16 building technologies driving the future*. Retrieved August 12, 2024, from https://openasset.com/blog/construction-technology/

Sreenath, S., Malik, H., Husnu, N., & Kalaichelavan, K. (2020). Assessment and use of unmanned aerial vehicle for civil structural health monitoring. *Procedia Computer Science, 170*, 656–663.

Stasiak-Betlejewska, R., & Potkány, M. (2015). Construction costs analysis and its importance to the economy. *Procedia Economics and Finance, 34*, 35–42.

Statista Research Department. (2019). *China's construction industry contribution shares to GDP 2018 to 2021*. Retrieved August 12, 2024, from https://www.statista.com/statistics/1068213/china-construction-industry-gdpcontributionshare/#:~:text=China's%20construction%20industry%20contribution%20share%20to%20GDP%202018%20to%202021&text=In%202018%2C%20the%20construction%20industry,a%20five%20percent%20annual%20growth

Stojanovska-Georgievska, L., Sandeva, I., Krleski, A., Spasevska, H., Ginovska, M., Panchevski, I., Ivanov, R., Arnal, I. P., Cerovsek, T., & Funtik, T. (2022). BIM in the center of digital transformation of the construction sector: The status of BIM adoption in North Macedonia. *Buildings, 12*(2), 218.

Struková, Z., & Líška, M. (2012). Application of automation and robotics in construction work execution. *AD ALTA: Journal of Interdisciplinary Research, 2*(2), 121–125.

Surbiryala, J., & Rong, C. (2019). Cloud computing: History and overview. In *2019 IEEE cloud summit* (pp. 1–7).

Takim, R., & Akintoye, A. (2002). Performance indicators for successful construction project performance. In D. Greenwood (Ed.), *18th annual Association of Researchers in construction management conference* (pp. 545–555). ARCOM, University of Northumbria.

Takyi-Annan, G. E., & Hong, Z. (2023). A bibliometric analysis of building information modelling implementation barriers in the developing world using an interpretive structural modelling approach. *Heliyon, 9*(8), e18601.

Thomas, P. Y. (2010). A potential paradigm for practising the scholarship of teaching and learning. *Cloud Computing, 29*, 214–224.

Trimble. (2023). *The rise of BIM in the UK*. Retrieved December 9, 2024, from https://www.viewpoint.com/en-gb/blog/the-rise-of-bim-in-the-uk?redirected=y#:~:text=The%202020%20BIM%20Report%20shows,compared%20with%201%25%20in%202020

TST Europe. (2023). *Construction industry statistics in the U.S. in 2023*. Retrieved December 9, 2024, from https://tst-europe.com/insights/construction-industry-statistics/#:~:text=Approximately%204.8%25%20of%20the%20entire,the%20construction%20industry%20was%204.4%25

Tucker, R. L., Haas, C. T., Glover, R. W., Alemany, C. H., Carley, L. A., Eickmann, J. A., Rodríguez, A. M., & Shields, D. (1999). Craft workers' experiences with and attitudes towards multiskilling. In: *Report no. 3, Center for Construction Industry Studies*, the University of Texas at Austin, Austin, TX.

Urie, M. (2019). *The Internet of things in construction*. Retrieved August 12, 2024, from https://cdn2.assets-servd.host/gt-marketintel/production/uploads/Articles/2019/IoT-in-Construction/1901_IoT-in-Construction.pdf

Vaduva-Sahhanoglu, A., Calbureanu-Popescu, M. X., & Smid, S. (2016). Automated and robotic construction: A solution for the social challenges of the construction sector. *Revista de Ştiinţe Politice, 50*, 1–11.

Van Tam, N., Quoc Toan, N., & Van Phong, V. (2024). Investigating potential barriers to construction digitalization in emerging economies: A study in Vietnam. *International Journal of Information Management Data Insights, 4*, 1–11.

Vitharana, V. H. P., De Silva, G. H. M. J., & De Silva, S. (2015). Health hazards, risk and safety practices in construction sites—A review study. *Engineer: Journal of the Institution of Engineers, 48*(3), 35–44.

Waal-Montgomery, M. D. (2015). *World's data volume to grow 40% per year & 50 times by 2020: Aureus*. Retrieved August 12, 2024, from https://e27.co/worlds-data-volume-to-grow-40-per-year-50-times-by-2020-aureus-20150115-2/

Wallen, M. C., & Sodergren, F. (2021). *Digital transformation through collaboration*. UMEA University.

Wang, P., Wu, P., Wang, J., Chi, H. L., & Wang, X. (2018). A critical review of the use of virtual reality in construction engineering education and training. *International Journal of Environmental Research and Public Health, 15*(6), 1204–1222.

Wetangula, J., & Mazurewicz, M. (2017). *Market research preliminary report: The construction market in Kenya*. Polish Investment and Trade Agency—Nairobi office. Retrieved December 9, 2024, from www.paih.gov.pl

Windapo, A. O. (2016). Skilled labour supply in the south African construction industry: The nexus between certification, quality of work output and shortages. *SA Journal of Human Resource Management, 14*(1), a750.

World Economic Forum. (2016). *Shaping the future of construction: A breakthrough in mindset and technology*. Retrieved December 9, 2024, from https://www.weforum.org/reports/shaping-the-future-of-construction-a-breakthrough-in-mindset-and-technology

Wu, P., Wang, J., & Wang, X. (2016). A critical review of the use of 3-D printing in the construction industry. *Automation in Construction, 68*, 21–31.

Yahya, M. Y. B., Hui, Y. L., Yassin, A. B. M., Omar, R., Robin, R. O., & Kasim, N. (2019). The challenges of the implementation of construction robotics technologies in the construction. In: *Proceedings of the international conference on built environment and Engineering*, Malaysia, 29–30 October 2018 (pp. 1–5).

Zaia, Y. Y., Adam, S. M., & Abdulrahman, F. H. (2023). Investigating BIM level in the Iraqi construction industry. *Ain Shams Engineering Journal, 14*(3), 101881.

Zhang, W. (2024). *Number of construction employees in China 2022, by region*. Retrieved December 9, 2024, from https://www.statista.com/statistics/279274/number-of-construction-employees-in-china-by-region/

Zhou, Z., Irizarry, J., & Lu, Y. (2018). Multidimensional framework for unmanned aerial system applications in construction project management. *Journal of Management in Engineering, 34*(3), 1–15.

Zingoni, T. (2020). Deconstructing South Africa's construction industry performance. *Mail and Guardian*. Retrieved August 12, 2024, from https://mg.co.za/opinion/2020-10-19-deconstructing-south-africas-construction-industry-performance/#:~:text=The%20South%20African%20construction%20sector,for%20construction%20in%20the%20country

Chapter 4
Sustainable Construction Practices in the Digital Era

Introduction

Sustainability plays an important role in the in the creation of a future that coming generations can inherit. Construction, which is a key driver of socio-economic and infrastructure development, sits at the forefront of this sustainable future (Oke et al., 2015). As such, delivering sustainable construction (SC) has become a topical issue among academics, researchers, industry practitioners and relevant stakeholders. This approach to construction has been described as designing, constructing and operating construction projects in an environmentally responsible way, including recycling and reusing materials while ensuring waste is reduced and energy efficiency is optimised (Stanca, 2023). As a result, studies have continued to explore the impact of SC on construction and the industry at large, as well as the factors that can promote this concept within the construction industry (Masyhur et al., 2024; Thounaojam et al., 2023). One emerging factor is the use of emerging digital technologies. As such, this chapter explores the concept of SC and the available practices. It also gives insight into the application of SC and the challenges hindering its effective adoption in diverse construction industries worldwide. The chapter further identified the different digital technologies for SC and the sustainability benefits of adopting these digital technologies.

Sustainable Construction

Traditionally, the construction industry, responsible for delivering public and private construction projects, has been described as a critical contributor to CO_2 emissions, high energy consumption, and an unsustainable environment due to the industry's activities. Rapid depletion of natural resources and pollution of natural habitats have become common due to industry activities (Ametepey & Aigbavboa, 2014; Du Plessis, 2002). Globally, the construction industry, in its delivery of

O. Oladinrin et al., *Ethical Practices for Sustainable Construction Digitalisation*, https://doi.org/10.1007/978-3-032-00151-1_4

construction projects, accounts for between 35 to 40% of the global energy consumption expended majorly around heating and cooling, with a possible increase of up to 50% if the energy spent on construction and demolition of AEC projects are taken into account. The industry further contributes to 39% of CO_2 emissions from the production of materials, construction and use of its products, as well as high wastage of resources (International Energy Agency, 2019; United Nations Environment Programme, 2020; Vasilca et al., 2021; Weniger et al., 2023; World Green Building Council, 2019). Ultimately, this high use of energy and production of CO_2, coupled with other unsustainable construction industry activities, has become a source of concern for practitioners, researchers and policymakers worldwide (World Green Building Council, 2019). This has necessitated the clamour for SC in recent times (Aghimien et al., 2019).

Brundtland Report (1987) defined sustainable development as "development that meets the needs of the present without compromising the ability of future generations to meet their own needs" (p. 43). Oke et al. (2015) noted that SC is crucial to achieving this development for the present and the future. SC has been described as the practice whereby the design, construction, and operation of buildings and other infrastructure are done in a manner that is environmentally responsible, resource-efficient and does not negatively impact society and the economy (Xero, 2024). Oke et al. (2015) described SC as a process that finds a balance between economic, environmental and social factors in building design, construction, use and maintenance. Stanca (2023) stated that SC is a design approach that reduces the impact of the construction process on the environment during the design stage and the whole life cycle of a construction project. As such, Ershadi and Goodarzi (2021) noted that SC requires the adopting of a long-term perspective of a building or infrastructure. Environmentally friendly buildings can reduce overall energy consumption and increase the availability of renewable energy. Compliance Chain (2024) also noted that SC is a veritable approach towards preserving the ecosystem, conserving natural resources and reducing energy consumption. Moreover, SC practices are essential for reducing the negative environmental impact of the construction industry, which includes high energy consumption, waste generation, and pollution. SC aims to offer lasting affordability, quality, and efficiency, thereby delivering value to clients and users, reducing adverse environmental impacts and enhancing economic sustainability (Bal et al., 2013). Azazga (2019) noted that SC leads to economic growth brought about by reduced energy consumption and improved efficiency, while Abdullah et al. (2021) noted that SC enhances building productivity and durability.

In attaining SC, studies have proposed green building materials characterised by their ability to offer sufficient performance and resilience with very little maintenance required. Additionally, they contribute to minimising the processing of raw materials, reducing pollution generated throughout the manufacturing and usage processes, and reducing energy consumption. These materials have significant potential for reuse and resource recovery, underscoring their importance in construction. Aside from reducing pollution, green building materials contribute to sustainable development (Aigbavboa et al., 2017; Bal et al., 2013; Patel & Patel, 2021).

Abera (2024) noted that one of the most essential strategies for achieving environmentally friendly construction is extensively using eco-friendly and sustainable building materials. Sustainable building materials promote the construction of more durable, healthier, and energy-efficient buildings besides countering environmental challenges. Using materials such as bamboo, recycled glass, hempcrete, recycled metal, recovered wood and algae-based materials to reduce the ecological impact of construction and promote sustainability is highly beneficial to attaining SC.

Employing SC techniques and materials can lead to longer-lasting structures, thereby reducing the need for frequent maintenance. This approach not only promotes environmental stewardship but also minimises lifecycle costs, boosting the project's economic viability (Abdullah et al., 2021). Modern materials, such as recycled polymers, green composites, and high-performance concrete, offer improved structural qualities while embodying sustainable development principles (Nwokediegwu et al., 2024).

Overview of Sustainable Practices in the Construction Industry

SC practices have continued to gain prominence in the global construction industry. Ranging from the use of sustainable materials, initiatives to reduce waste on construction projects, adopting renewable energy and energy efficient materials and design approaches, protecting the environment and ensuring indoor comfort, SC has continued to be a success criterion in many construction projects (Kaiv, 2023). Bal et al. (2013) further noted that certain sustainability features must be evident for a construction project to be termed sustainable. This includes designing structures that will last the expected lifespan, adhering to sustainability standards and regulations, considering both the short and long-term impact of the project on the environment and prioritising stakeholders' contribution in the delivery of the project.

Evidence of the implementation of sustainable construction practices has continued to emerge over the years. Archistar (2024) noted that in Asia-Pacific regions, there is an adoption rate of about 43% for green construction. New buildings boast about a 63% adoption rate for green construction, while urbanised areas, including Australia, Japan, Hong Kong and Singapore, are at the forefront of green building implementation. It was further noted that Brisbane, Melbourne, Perth, Canberra, and Sydney, Australia, rank among the places where green construction has been implemented. In China, the construction industry's high energy consumption and waste generation led to the enactment of a series of environmental policies to encourage the use of green building materials and technologies. These policies cut across solid waste pollution issues, response to climate change, standards for green construction, and protection of the environment (Cheng et al., 2016; Yang et al., 2021). In Brazil, SC achieved through green technology is rapidly gaining recognition, with at least 3500 certified green buildings recorded in 2023 (Amaral, 2023).

In the United States of America (USA), it has been noted that one of the major factors that have led to the rise of SC is the widespread adoption of green building certification systems such as Leadership in Energy and Environmental Design (LEED) and the WELL Building Standard (Level Construction Recruitment, 2024). This case is similar in the United Kingdom (UK), where several building regulations are designed to shape insulation levels and other sustainability requirements in buildings. This is coupled with the Building Research Establishment Environmental Assessment Method (BREEAM) developed by the Building Research Establishment and the UK's net zero carbon building standards (Rockfon, 2023). These systems have helped shape the widespread adoption of SC, which is experienced in the USA and the UK.

Challenges Facing Sustainable Construction

Despite this implementation of sustainable practices globally, the construction industry in many countries, particularly developing ones, is still lagging in integrating sustainable principles in their project delivery. This is the case as evident in studies conducted in Ethiopia (Baron & Donath, 2016), Ghana (Djokoto et al., 2014), Kuwait (Al-Sanad, 2015), Malaysia (Masyhur et al., 2024), Nigeria (Eze et al., 2023), South Africa (Saad, 2016), Vietnam (Nguyen et al., 2017) and Zambia (Oke et al., 2019). In Malaysia, Masyhur et al. (2024) noted that despite accounting for 25% of greenhouse gas emissions and 40% of energy consumption, the construction industry is still lagging in implementing SC practices. Ironically, Malaysia has long-standing environmental policies, yet the use of green practices that will protect the environment and its resources is still very low with in the construction industry. In Nigeria, Eze et al. (2023) noted that SC practices are very low owing to several factors, including resistance to change, absence of SC regulation, lack of funding for research and development, perceived cost implication, lack of incentives to adopt SC practices, absence of reliable suppliers and green building experts. In a similar study, Aghimien et al. (2019) discovered that resistance to change, client preference, fear of the increased cost of investment and inadequate knowledge and understanding of SC are hindering the implementation of SC practices in Nigeria and South Africa. In Zambia, Oke et al. (2019) observed that while there is a growing awareness of SC in the country's construction industry, actual implementation is still low. Issues around fear of higher investment costs, absence of local green certification, lack of government policies, and lack of financial incentives were noted as key hindrances to the implementation of SC.

In Vietnam, the story is the same as Nguyen et al. (2017) noted the slow implementation of SC due to several issues, including slow and ineffective administration process in policy-making, absence of a comprehensive policy package to guide action on SC, absence of explicit financing mechanism and inadequate fiscal incentives. In Indian, Gehlot and Shrivastava (2022) noted that the absence of a skilled SC workforce, resistance from clients and the lack of awareness of SC practices within the construction industry are deterring the SC implementation. Bora et al. (2023) further noted that although three green building rating systems exist in India, there is still no systematic regulatory framework designed to effectively incorporate sustainability principles in the Indian construction industry. The implication of this is the slow pace of adoption of SC practices in the country's construction industry. Despite this slow implementation, Thounaojam et al. (2023) highlighted some common SC practices used in the Indian Construction Industry. These include safety inspection, safety training, and education, as well as concerted efforts in preserving archaeological sites, vegetation, and trees, ensuring effective machinery handling and utilisation, and quality management systems. In Pakistan, Azeem et al. (2020) noted that despite having several sustainability-related professional bodies, the implementation of SC in the country's construction industry is still challenged by poor awareness regarding the inherent benefits of employing SC practices, lack of incentives from government, and absence of green building codes and regulations.

Evidently, the implementation of green and sustainable practices is challenged by several issues. These issues can be summarised as:

1. Absence of explicit financing mechanism.
2. Absence of local green certification,
3. Absence of reliable suppliers and green building experts.
4. Absence of SC regulation.
5. Client's preference/lack of interest.
6. Inadequate knowledge and understanding of SC.
7. Lack of awareness of SC practices.
8. Lack of funding for research and development.
9. Lack of government policies to guide action on SC.
10. Lack of incentives to adopt SC practices.
11. Misconception about SC.
12. Perceived cost implication.
13. Poor enforcement of policies.
14. Poor communication with stakeholders.
15. Resistance to change among professionals.
16. Slow and ineffective administration process in policy-making.

Sustainable Construction Practices

SC practices adopted in building construction vary across different countries, as are the measures adopted in their implementation. On a broader spectrum, SC principles are rooted in sustainability's environmental, social and economic pillars. However, diverse SC practices shaped by these principles have continued to emerge over the years. Similarly, from the perspective of construction organisations, diverse management practices can also be incorporated to ensure SC is achieved. Tan et al. (2011) proposed six practices from an organisational perspective: compliance with

SC legislations, adopting green designs and supply chains, promoting technology and innovation, re-organisation of structure and processes within the organisation, promoting education and training and developing a clear measurement and report tool for SC. Kibert (2013) also proposed some practices from the sustainability paradigm shift. This includes construction ecology, design for the environment, biomimicry, ecological economics, life cycle assessment and life cycle costing. Additionally, prefabrication and modular construction, value engineering and management, lean construction, and other emerging digital technologies such as robotics and automation (including 3D printing, robots, and drones) have been noted to be valuable management principles that organisations can use to promote their quest for SC (Li et al., 2019; Oke et al., 2019).

Energy Efficiency Studies have continued to clamour for the need for energy efficiency in buildings. This is because energy efficiency is integral to green and sustainable buildings (Shi et al., 2016). As such, studies have continued to emphasise the need for energy efficiency in buildings to reduce CO_2 emissions, limit energy consumption, and improve overall building performance (Chen et al., 2015; Hafez et al., 2023). Energy efficiency initiatives such as design and building orientation (passive solar design), insulation, LED lighting, high-efficiency HVAC systems and energy management processes have all been proposed for effective SC in buildings (Altan et al., 2015; Dobias & Macek, 2014; Papadakis & Katsaprakakis, 2023).

Sustainable Materials The type of materials used for construction is crucial to attaining SC. These materials are usually environmentally responsible and have little impact on the environment. Additionally, sustainable materials offer social and economic benefits along with environmental gains. They are readily available locally and meet the users' needs and health considerations. Economically, they are financially viable, allow cost savings on transportation due to their local availability, and are easy to maintain. From the environmental dimension, sustainable materials are thermally efficient, recyclable and reusable. Their production promotes using renewable resources, reducing waste and pollutant generation. Examples include cob, bamboo, straw bales, recycled steel, reclaimed wood, recycled plastics, rammed earth, earth blocks, and hempcrete, among others (Barbulianno, 2020; BigRentz, 2023; Green, 2023; Wood, 2024). Furthermore, emerging nature-inspired materials have continued to emerge. Bio-concrete and blocks, bioplastics, and photobioreactors are also useful SC materials that can drive energy efficiency, CO_2 reduction, and efficient release of oxygen and hydrogen to reduce air pollution, production of biofuel and efficient treatment of waste water (Elrayies, 2018; Hanafi, 2021; Yaman et al., 2021).

Circular Economy The circular economy has recently become a prominent concept in the quest for attaining SC. By embracing circular economy principles centred on reducing, reusing, recycling, recovering, remanufacturing, and redesigning across the lifecycle, the attainment of SC can be facilitated (Asante et al., 2022). A shift towards a circular economy is essential for realising sustainable development goals and alleviating the adverse effects of conventional linear economic models. This transition offers manifold advantages, including diminished resource utilisation and waste production, employment opportunities, and curbing greenhouse gas emissions and pollution. (McBride Sustainability, 2023). Diverse studies have demonstrated the global inadequacy of the traditional linear economy, characterised by the 'take-make-dispose' model, in sustaining economic growth, social well-being, and environmental health (Wijewansha et al., 2021). Hence, the circular economy has been recognised as a more regenerative and restorative alternative (Kern et al., 2019). Circular economy has been described as a paradigm designed to minimise waste, optimise material value throughout production, reduce pollution, mitigate resource depletion, enhance material durability, prolong the value of materials within the value chain, bolster sustainability, foster collaboration, and efficiently manage resources to yield environmental, social, and economic benefits (Kazancoglu et al., 2020). This makes this concept a crucial approach towards attaining SC.

Nature-Based Solutions Nature-based solutions are actions that effectively and adaptively address social, economic and environmental challenges while simultaneously protecting, conserving, restoring and sustainably managing natural or modified terrestrial, freshwater, coastal and marine ecosystems. These actions are designed to improve well-being, ecosystem services, resilience and biodiversity (United Nations Environment Assembly, 2020). International Union for Conservation of Nature (IUCN) (2020) described it as an innovative approach towards conserving nature while protecting, managing and restoring the environment and, in the process, delivering benefits that are tangible and sustainable for the people. Nature-based solution offers prospects for urban development resilience to climate change by simultaneously addressing various sustainability challenges (Szkop et al., 2021). They encompass a variety of methods, including sustainable infrastructure and adaptation based on ecosystems, which use nature to strengthen the sustainability and resilience of urban environments (Palomo et al., 2021; Wickenberg et al., 2021).

Studies have emerged on the benefits of deploying nature-based solutions in attaining SC within the construction industry. Nature-based solutions deliver long-term environmental and social benefits, with the potential to offset initial high costs and generate net gains. Nature-based solutions such as green infrastructure and ecosystem-based adaptation, which leverage natural processes to transform urban environments into more sustainable and resilient spaces, can deliver SC. Moreover, nature-based solutions offer social, economic, and ecological advantages and are crucial in lowering the impacts of climate change and enhancing human welfare (Cohen-Shacham et al., 2016; Eggermont et al., 2018; Palomo et al., 2021; Wickenberg et al., 2021).

Biomimicry The concept of biomimicry, a nature-based solution, has been described as the practice of learning from nature's strategies in creating sustainable products (Biomimicry Institute, 2024). It follows the principles of emulating, ethos and reconnecting to deliver sustainable products and technologies that are beneficial to man and nature (Oguntona & Aigbavboa, 2017). Al-Amin and Taleb (2016) noted that using nature, biomimicry will enable a more sustainable future by solving the earth's problems caused by overexploitation of natural resources, erratic population growth and industrialisation.

Ecological Economics This transdisciplinary field explored the correlation between the economy and the environment. According to Neo (2009), ecological economics cuts across valuing natural resources, ecologically integrated technology, and sustainable agriculture and development. Capra and Jakobsen (2017) noted that the ecological economy is an economic practice that explores the economy as part of society and the environment instead of separately. Oke et al. (2018) submitted that an ecological economy offers long-term cost benefits when implemented correctly. This makes it a suitable practice for attaining SC, particularly from the economic dimension (Kibert, 2013).

Lean Construction The concept of lean in construction project delivery aligns with SC as it offers veritable approach towards managing waste on construction projects (Adegbembo et al., 2016). Studies have also noted that lean construction offers value for clients, reduces wastage and promises better overall project delivery when used correctly (Freitag et al., 2020; Mamillapalli et al., 2021). As such, it can be deduced that through lean thinking, SC can be achieved within the construction industry by optimising resource utilisation, waste reduction and prioritisation of safety (Nahmens & Ikuma, 2011; Oke et al., 2019).

Value Management Oke et al. (2015) proposed value management as an option for attaining sustainability, particularly in the economic dimension. This is because both sustainability and value management are proponents of achieving value for money. Noor et al. (2015) have earlier noted that VM has been promoted in the Malaysian construction industry as a valuable strategic planning tool suitable for the delivery of SC. Similarly, Yekinni et al. (2015) observed a relationship between VM and sustainability in design and production, as they both lead to achieving the best value in terms of quality and cost. Oke et al. (2015) noted that the VM team strives to analyse the function through a VM workshop and look for better alternatives. This process helps to identify elements with unnecessary cost implications that might harm value. Abidin and Ijias (2006) submitted that applying value management can assist construction organisations in effective decision-making, and it is important to incorporate sustainability issues into construction projects.

Modular Construction Modular construction—an off-site construction approach—is a sustainable approach to delivering construction projects. Studies have compared the traditional and modular construction methods and have favoured the latter due to its ability to solve crucial problems associated with the conventional approach (Azman et al., 2012; Karthikeyan et al., 2018; Subramanya et al., 2020). The modular construction system for project delivery has been described using terms such as prefabrication (Li et al., 2018), modular integrated construction (Abdelmageed et al., 2020; Wuni & Shen, 2020), off-site manufacturing/construction (Nawi et al., 2019; Wang et al., 2020), the industrialised building system (Ayoola & Aghimien, 2017; Nawi et al., 2019) as well as modern method of construction (Musa et al., 2014; Nawi et al., 2019). Modular construction is an old approach, but technological advancements, the changing economic climate, and the ever-changing tastes of construction clients have attracted an unparalleled quest for this system in modern-day construction (Bertram et al., 2019).

Modular construction has been described as a process involving the prefabrication of modules with up to 95% completion (including finishes) in an off-site factory (Wang et al., 2020). Musa et al. (2014) noted that the modules are mass-produced using specified materials and standards to increase construction speed. These modules are then transported to the site and assembled based on the design and specification with minimal effort (Hamid et al., 2008). As a result, one of the major benefits of modular construction is its ability to provide timely delivery of projects (Bendi et al., 2021). More so, modular construction offers cost savings and SC by reducing the environmental impact of projects and waste generation, reduction in the industry's over-reliance on on-site skilled labour, better safety on construction sites, better quality delivery of projects, saving spaces, among others (Aladeloba et al., 2015; Čuš-Babič et al., 2014; Jin et al., 2020; Musa et al., 2014; Wasim et al., 2020; Wuni & Shen, 2020). Due to these stated benefits, modular construction has been receiving significant attention in both developed and developing nations such as the USA, UK, Germany, China, Malaysia, Australia, Japan, Sweden, and Singapore (Ferdous et al., 2019; Kamali & Hewage, 2016; Khalil et al., 2016; Nawi et al., 2019; Wuni & Shen, 2020; Xu et al., 2019).

Emerging Technology, Digitalisation and Automation Technological advancement offers SC opportunities. For instance, advancement in nanotechnology can help reduce carbon footprint while improving energy efficiency and reducing waste (Farhadi, 2023). Nanotechnology, which is the process of using atoms and molecules to create new materials, has continued to gain prominence as a viable SC technology (Babuka, 2016; Oke et al., 2017). Moreover, digital technologies such as Building Information Modelling (BIM) allow detailed representation of diverse elements of a structure, which is very useful in reducing waste and rework during construction (Eze et al., 2022). Studies have noted that BIM offers better quality and efficient productivity in construction project delivery while eliminating increased costs associated with rework when clashes are detected in designs. It can aid in understanding expected energy usage and carbon footprint monitoring, which is essential to SC (Davies et al., 2018; Olawumi & Chan, 2018). Similarly, robotics and automation such as autonomous robots, 3D printing and drones remove workers from harmful situations on site as these machines can be used instead. This allows workers to manage the operation of those machines from a safe location off-site, thereby ensuring their safety and promoting social sustainability (Aghimien et al., 2020).

To summarise, SC practices for building construction and for organisations seeking to achieve SC include but are not limited to:

1. Adopting green designs and supply chain.
2. Automation in construction.
3. Biomimicry.
4. Circular economy.
5. Clear measurement and report tool for SC.
6. Compliance with SC legislation.
7. Construction ecology.
8. Design for environment.
9. Designing buildings that maximise natural light, heat, and energy efficiency.
10. Ecological economics.
11. Enhancing indoor environmental quality (IEQ).
12. Environmental impact assessment.
13. Green roof and living wall.
14. Integrating smart technology in buildings/digitalisation.
15. Integration of renewable energy sources in construction.
16. Lean construction.
17. Life cycle assessment and costing.
18. Lifecycle assessment.
19. Nature-based solutions.
20. Prefabrication and modular construction.
21. Prioritising waste reduction.
22. Promoting education and training and developing.
23. Promoting energy efficiency in design.
24. Promoting technology and innovation.
25. Promoting water efficiency.
26. Re-organisation of structure and processes within the organisation.
27. Stakeholder management.
28. Sustainability audits and certification.
29. Use of green roofs and walls.
30. Use of sustainable and green materials.
31. Value engineering and management.

Digitalisation and Sustainable Construction

World Economic Forum (2022a) noted some emerging technologies that can drive the reduction in CO_2 emission generated by the construction industry and, by extension, lead to achieving SC. These include artificial intelligence (AI) and machine learning, the Internet of Things (IoT), digital twin, big data analytics (BDA), drones, automation and robotics, augmented and virtual realities, blockchain and cloud computing. This technology offers many benefits to construction, including reducing waste, reducing carbon footprint and improving energy efficiency (Adames, 2023). Lu et al. (2024), while exploring digital technologies for construction sustainability, concluded that BIM, IoT, big data and AI are all crucial to attaining SC. It was further noted that these emerging technologies can be applied in construction to attain sustainability in different ways, including integration, ensuring efficient collaboration, optimising construction processes, simulation, effective decision-making, easy monitoring and control, and effective training of workers. In exploring digital technologies for sustainable buildings, Asif et al. (2024) noted that digital technologies, when integrated into construction, can offer better integration, flexibility, efficiency in energy consumption, resource optimisation, ensure the operational process is efficient while minimising the adverse impact of construction on the environment. Moreover, Adames (2023) noted that emerging technologies make it easier to select suitable green materials for construction and energy use in buildings.

Barnes (2023) submitted that through digital technologies, construction organisations could reduce the CO_2 emissions on their project, improve their waste management, design energy-efficient buildings, and ensure effective monitoring of their resources. Digital technologies offer effective data capture on construction projects from inception to completion, which aids decision-making. Moreover, IoT and sensors can be used to track resources in real-time and monitor resource consumption on projects to avoid wastage (Barnes, 2023; Lu et al., 2024). Through technology, paper use on projects is reduced, and collaboration can be done digitally (Shuja, 2024), while decision-making is easier, more efficient and data-driven through BDA (Fern, 2024). Build Chain (2024) also noted that digital technologies are making construction procurement more efficient. Through digitalising the procurement system, organisations can improve the supply chain while reducing the risk of acquiring materials that are not needed, allowing easy access to environmentally friendly materials. Moreover, technologies like blockchain offer transparency in the supply chain by offering traceable and transparent transactions, allowing materials to be acquired ethically and environmental claims to be easily verified. Evidently, digital technology deployment in construction offers invaluable benefits for sustainability in construction. The benefits can be summarised as:

1. Better monitoring and control of resources.
2. Easy selection of suitable green material.
3. Effective decision-making,
4. Effective training of workers.
5. Effective waste management.
6. Efficiency in the supply chain.
7. Efficient collaboration.
8. Efficient procurement system.
9. Efficient use of resources.
10. Energy efficiency.
11. Flexibility in construction projects.
12. Minimising the impact of construction on the environment.

13. Optimising construction processes,
14. Reduction in CO_2 emission/carbon footprint.
15. Smart monitoring of performance of buildings.

Understanding Digitalisation and Sustainable Construction from the Perspective of Past Studies

To further shed light on the relationship between sustainable construction and digitalisation, past studies were explored using a bibliometric approach, which is a systematic analysis of literature published within a selected area through the use of quantitative tools and map creation (Hallinger & Kovačević, 2023). Bibliographic data were gathered from the Scopus database to conduct the bibliometric review. The search for related literature was conducted using carefully selected keywords and two Boolean operators ("OR", "AND"). The search protocol followed: TITLE ("Sustainable construction" OR "Sustainability" OR "Green construction" OR "Green building" AND TITLE-ABS-KEY ("Digital technology" OR "digitalisation" OR "Digital transformation" OR "smart technologies" OR "Artificial intelligence" OR "Internet of things" OR "IoT" OR "Building Information Modelling" OR "BIM" OR "Big data analytics" OR "drones" OR "Unmanned aerial vehicles" OR "UAVs" OR "Sensors" OR "Robotics" OR "Autonomous robots" OR "Additive manufacturing" OR "AI" OR "3d printing" OR "cloud computing" OR "blockchain" OR "digital twin"). The search was limited to 2014 to 2024 to capture documents published in the last 10 years to gain recency. The initial search revealed 2537 documents. To ensure only relevant papers related to construction are extracted, the result was further refined in the subject area (engineering), source type (journal or conference proceedings), and language (English). Based on these new parameters and screening for duplication or incomplete details, 775 documents emerged and were used for the bibliometric analysis.

The overall information of the bibliographic data revealed that the 775 documents contain 490 articles and 285 conference papers, all published in 403 different sources. Regarding frequency of publication, there has been a steady increase in the number of publications per year since 2014, as seen in Fig. 4.1. The year 2024 recorded the highest number of documents on sustainability and digitalisation, with 198 documents. This implies an increased interest in this topic and its importance to sustainability in construction.

Further observation of the countries of publication revealed that the extracted documents came from 63 different countries, with China having the highest number of publications (91) with a total citation of 244. Other prominent countries researching in this area include India (53 documents; 786 citations), Italy (35 documents; 968 citations), the UK (35 documents; 2942 citations), the USA (22 documents; 334 citations) and Malaysia (21 documents; 275 citations). This implies that this area of research has received attention from diverse parts of the world, a further testament to its importance.

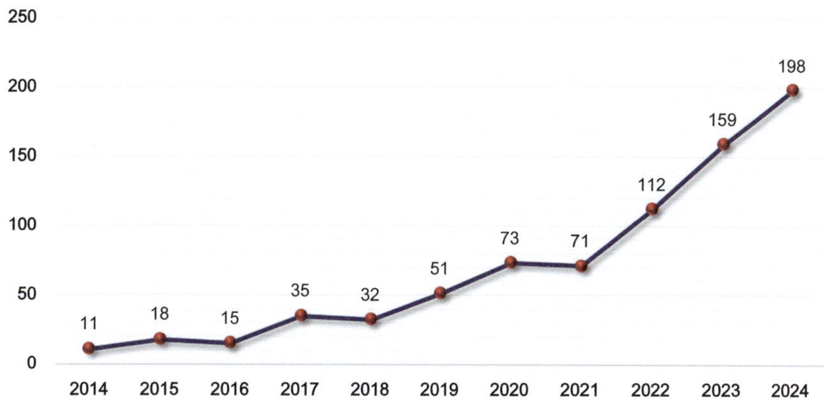

Fig. 4.1 Publication per year of digitalisation studies and sustainability studies

The most impactful documents in this area of research based on their citation index were further explored. The result revealed that the work of Ford and Despeisse (2016), which explored the advantages and challenges of 3D printing and sustainability, was the most cited document with 1289 citations. This article was published by the Journal of Cleaner Production. This is followed by Bibri (2018) which proposed an analytic framework for using sensor-based big data in a quest for a smart, sustainable city of the future achieved through IoT, and published by Sustainable Cities and Society journal. This study has been cited 533 times. Also, Bai and Sarkis (2020), Hager et al. (2016), Joshi and Sheikh (2015), and Lu et al. (2017) have all contributed significantly, with over 400 citations each. While Joshi and Sheikh (2015) explored 3D printing in aerospace by concentrating on its long-term impact on sustainability, Bai and Sarkis (2020) beamed their searchlight on blockchain technology for supply chain transparency and sustainability. The study of Lu et al. (2017) was a review of BIM in green buildings, while Hager et al. (2016) explored 3D printing as the future of sustainable construction.

In the quest to determine the major areas of focus of these researches, a word cloud of the top 50 terms in the extracted document was analysed using Biblioshiny in R-Studio and presented in Fig. 4.2. The analysis revealed that common terms were sustainable development ($f = 424$), architectural design ($f = 140$), construction industry ($f = 91$), IoT ($f = 91$), energy utilisation ($f = 82$), life cycle ($f = 82$), decision making ($f = 72$), green buildings ($f = 70$), BIM ($f = 66$), environmental impact ($f = 64$), and AI ($f = 50$). These top common terms cut across SC and emerging technologies.

Further analysis of common terms was done using the visualisation of similarities viewer (VOSviewer) software. This software allows a network visualisation of the most co-occurring terms. The analysis revealed that the 775 documents earlier extracted had 5396 common terms. Setting the minimum co-occurrence at 10, a total of 134 terms clustered into five groups, as seen in Fig. 4.3. These five clusters

Fig. 4.2 Word cloud of common terms in digitalisation and sustainability studies

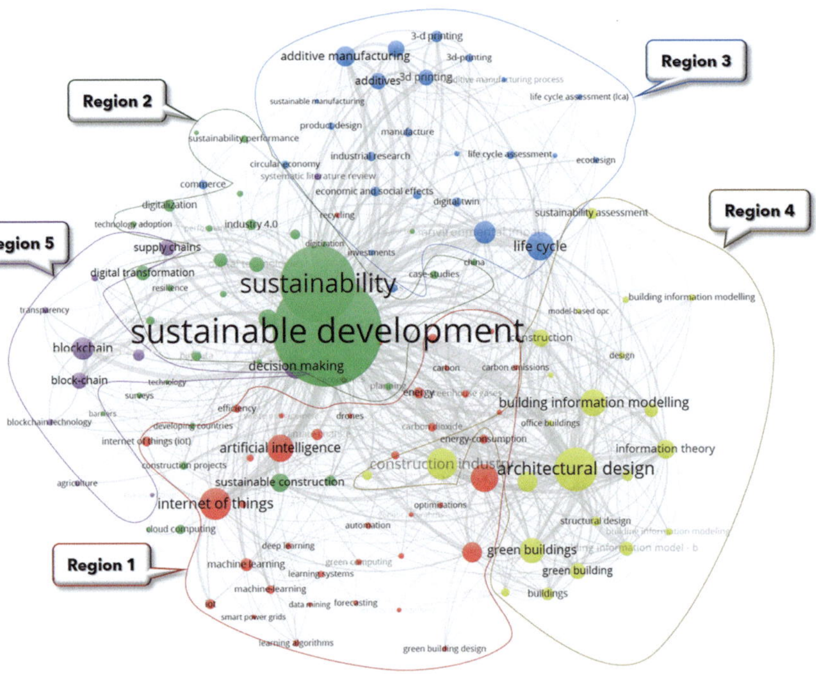

Fig. 4.3 Network visualisation of co-occurring terms in digitalisation and sustainability studies

had 4002 links to one another and 13,578 total link strength (TLS). The five clusters are presented below.

Cluster 1: AI and IoT for energy efficiency and sustainability in buildings. The first cluster in Region 1 on the map has 40 co-occurring terms. The major terms in this cluster are IoT (95 occurrences, 102 links), AI (78 occurrences, 105 links), energy utilisation (82 occurrences, 111 links) and energy efficiency (55 occurrences, 97 links). These key terms show studies exploring AI, IoT and machine learning applications in efficient energy consumption in buildings. Other prominent terms are deep learning, learning systems, automation, data mining, learning algorithms, smart power grids, green building design, energy, greenhouse gases, and carbon missions, among others. Based on the latent similarity of these terms, this cluster was named '*AI and IoT for energy efficiency and sustainability in buildings*.'

Cluster 2: Sustainable development through the adoption of digital technologies. The second cluster in Region 2 on the map revealed 34 co-occurring terms. Prominent terms in this cluster are sustainable development (425 occurrences, 132 links) and sustainability (273 occurrences, 129 links). These two terms serve as the focal point upon which other clusters relate. For instance, the four major terms in cluster 1 (AI, IoT, energy utilisation and energy efficiency) all have a strong link strength of above 35 with sustainable development. This shows that there has been significant mention of these terms in relation to sustainable development and sustainability. Other key terms in this cluster are digital technologies, sustainable construction, environmental sustainability, social sustainability, economic sustainability, cloud computing, technology, big data, digital transformation, technology adoption, industry 4.0, digitalisation, innovation, sustainability performance, sustainable development goals. These key terms relate to general studies exploring '*sustainable development through the adoption of digital technologies*.'

Cluster 3: Additive manufacturing and its impact on construction life cycle and the environment. The third cluster in Region 3 on the map revealed 25 co-occurring terms. Lifecycle (84 occurrences, 111 links), environmental impacts (63 occurrences, 111 links), additive manufacturing (56 occurrences, 65 links), 3D printing (41 occurrences, 69 links), 3D printers (45 occurrences, 65 links), and additives (43 occurrences, 67 links) are the most prominent terms in this cluster. Other key terms include digital twin, manufacturing, costs, investments, additive manufacturing process, lifecycle assessment, eco-design, and circular economy, among others. This cluster was subsequently named '*additive manufacturing and its impact on construction life cycle and the environment*.'

Cluster 4: BIM in sustainable design and construction. The last cluster in Region 4 on the map revealed 23 co-occurring terms. The major key terms in this cluster are architectural design (109 occurrences, 139 links), construction industry (93 occurrences, 106 links), green buildings (72 occurrences, 90 links), BIM (76 occurrences, 94 links), and information theory (50 occurrences, 77 links). In this cluster, diverse variants of BIM emerged, showcasing the high reference for BIM. Others include sustainability assessment, energy conservation, intelligent buildings, information management, buildings, construction, and decision support systems. Based on the latent similarity in these terms, this cluster was named '*BIM in sustainable design and construction*'.

Cluster 5: Sustainable supply chain through blockchain technology. The last cluster in Region 5 on the map revealed 12 co-occurring terms. Prominent among these terms are decision-making (64 occurrences, 113 links), blockchain (63 occurrences, 80 links), and supply chains (43 occurrences, 80 links). Others include blockchain, blockchain technology, risk assessment, supply chain management, and transparency. These clusters are evidence of studies focusing on '*sustainable supply chain through blockchain technology*'.

These five clusters revealed the significant attention garnered by digitalisation and sustainable construction over the last 10 years. AI, IoT, BIM, blockchain, and 3D printing are all features prominently in ensuring energy efficiency, reduction of negative environmental, impact, delivering sustainable design and construction, and ensuring efficient supply chain process.

Digital Technologies of Sustainable Construction

The importance of digital technologies to the delivery of SC has led to several studies unearthing the applicable emerging technologies that construction organisations can invest in for optimum sustainability benefits. Asif et al. (2024) noted key technologies that can benefit SC. These are BIM, Building Management Systems, smart meters, sensors, 3D printing, robotics, BDA, IoT, machine learning, AI, digital twin, blockchain, and cyber security. Barnes (2023) noted digital construction could be through BIM, virtual design and construction, cloud-based collaboration, IoT, augmented and virtual realities, robotics, and automation.

Building Information Modelling Barnes (2023) observed that by integrating digital technologies, SC can be achieved through the reduction of carbon footprint on the project. This is because digital technologies like BIM offer efficient collaboration and coordination that reduces energy spent on delivering a project and reduces carbon footprint throughout the construction lifecycle. Also, BIM offers optimisation of material usage, thereby reducing the waste generated on construction projects and the huge cost incurred due to this waste. Asif et al. (2024) also noted that BIM can be very useful in ensuring sustainability assessments in construction projects. This technology design accuracy, modelling of energy usage and lifecycle assessments even before construction, thereby allowing effective planning.

Internet of Things It has been observed that IoT can be very effective in effectively utilising materials as it allows easy tracking of materials and equipment on construction sites. Also, IoT allows proactive maintenance of buildings, which ultimately leads to these buildings serving their expected lifespan. Barnes (2023) noted that by embedding IoT-driven sensors in buildings, information on energy consumption, water usage and overall resource use can be gathered and used for effective management. This makes IoT very useful technology in real-time data collection for monitoring and control. Moreover, Radio Frequency Identification (RFID), which is among the most common IoT, has become prominent in ensuring effective logistics management and has been noted for its ability to contribute to the reduction of the carbon footprint of construction projects (Sajjad et al., 2023; Tao et al., 2018). Studies have also noted green microgrids can be activated via IoT when conventional non-renewable systems are inaccessible, supplying a sustainable energy source in critical circumstances. Paired with smart devices, IoT allows power systems to operate more effectively, resulting in lower energy consumption. By incorporating IoT devices and edge computing, the plant can gather and evaluate data in real-time, optimising production processes and reducing environmental impact (Da Silva, 2020; Wu et al., 2020).

Blockchain Technology Teh et al. (2020) noted that blockchain technology includes a digital ledger and a distributed network of peers to-peers that form a shared database. Saberi et al. (2020) noted that the uniqueness of this technology is its decentralisation, which includes a shift of control from a centralised entity to a distributed network; security, which is ensured through a record of transactions saved in several distributed nodes; audibility, which happens with the approval of the transaction validity by the majority of nodes; and intelligent execution since the processes can be executed by smart contracts. Since the construction industry's supply chain is complex regarding subcontracting and liabilities, transparent value transactions are required. Blockchain technology is a safe method of tracking materials (Wolf, 2022).

Blockchain has been noted for ensuring transparency and accountability in projects (Xu et al., 2023). Through this transparency, blockchain technology can further promote a circular economy through transparency in recycling and reducing waste on projects (Dagou & Gurgun, 2024; Shojaei & Naderi, 2024). Also, blockchain can help smoothen the decision-making process involved in efficient energy use by optimising environmental data management (Yang et al., 2023). Moreover, it offers an opportunity to verify records on waste disposal on construction projects. This information can then be used to improve disposal practices and promote efficient circularity on projects (Ma et al., 2024).

Additive Manufacturing Additive manufacturing, also known as 3D printing, reduces material wastage and allows sustainable materials, including recycled plastics and bio-based composites, to be used for construction project delivery (World Economic Forum, 2022b). Wang et al. (2020) noted that human errors in construction practices frequently result in damage to materials and less efficiency of energy in buildings. Waste from damaged or unused materials adds to the contamination of the environment over time. Traditional construction methods rely heavily on labour- and cost-intensive resources such as cement, formwork, and metal reinforcements, all of which have a significant carbon footprint and contribute to global warming. The construction industry also has higher injury and fatality rates than most other industries (Aghimien et al., 2020). 3D printing can help meet the demand for environmentally friendly solutions and lower CO_2 emissions. It offers greater design flexibility, increased automation, and a lower requirement for resources and capital (Sakin & Kiroglu, 2017). Studies have also noted that 3D printing offers fewer logistics in construction, can deliver more complex designs at affordable costs and can employ more local sustainable materials, which contributes to the sustainability of the project.

Automation and Robotics Automation and robotics have also continued to gain prominence in construction. Robots are being utilised to create complex structures using alternative construction materials. For automated circular construction, data on the materials stored in buildings can be collected, indexed, and tracked. To improve infrastructure maintenance, autonomous robotic technology is being explored for its ability to detect and address problems before they escalate. For instance, robots are established to continuously monitor potholes, allowing for timely repairs and preventing minor damages from raising major structural issues. This proactive strategy seeks to extend the life of infrastructure while minimising the environmental impact of unneeded maintenance or early replacements (Kiy, 2022). It has also been noted that robots can help enhance efficiency and sustainability. This is done through the use of robots for maintenance and inspections, which often increases the accuracy of maintenance data and promotes the safety of workers (Halder & Afsari, 2023; Lu et al., 2020).

Digital Twins Sacks et al. (2020) submitted that digital twins, which initially act as a data-driven model for handling and monitoring physical systems, have gained popularity in the last 10 years, especially for manufacturing, production, and operations. Digital twin is a replica of physical structures which can be used to monitor and analyse the performance of such structures. In construction, digital twin applications are mainly employed during the operation and maintenance, emphasising infrastructure assets. It has been observed that during construction, this technology can help allocate resources efficiently and effectively monitor construction in real time. By doing this, delays in projects and wastage of materials are avoided. Moreover, the virtual representation of the physical structure allows effective management through timely identification and maintenance of failing components (Long et al., 2024; Ngcobo et al., 2024).

Artificial Intelligence and Big Data Analytics AI and BDA can impact the construction industry by predicting the accessibility of reusable, recyclable, and waste materials using images (Wolf, 2022). This predictive capability is vital when applying reclaimed materials because available resources influence project design. AI supports the ecological component of sustainability by tracking energy consumption, waste production, utilisation of natural resources, and the overall environmental impact of construction operations. It can help the construction industry handle waste from demolition and construction, evaluate the air condition, and handle wastewater more efficiently. AI also improves energy management, tracks environmental impacts, and aids decarbonisation efforts (Lambert, 2022; Tan, 2024).

Summary

This chapter shed light on SC and its practices. The chapter concludes that while examples of the application of SC are evident in the body of literature, there is still a lag in adoption in many countries, particularly developing ones. This lag is due to several issues, including local green certification, regulations and green experts, lack of interest, poor awareness and knowledge, misconception of SC, and absence of incentives to adopt SC, among others. With the growing concern about the negative impact of the construction industry on the environment, employing SC practices is no longer an option but a necessity. Practices such as utilising green technologies and materials, energy-efficient designs, water conservation, circular economy, nature-based solutions, biomimicry, ecological economics, lean construction, value management, smart technologies, and modular construction, among others, were unearthed as valuable SC practices. When digital technologies are implemented correctly, SC practices can lead to CO_2 reduction, efficient energy consumption, a transparent and reliable supply chain, reduced wastage, effective monitoring and control, and effective decision-making, among others. These technologies include BIM, 3D printing, sensors, drones, automation, robotics, BDA, IoT, machine learning, AI, digital twin, cloud computing, augmented and virtual realties, and blockchain.

References

Abdelmageed, S. M., Abdelkhalek, S., & Zayed, T. (2020). Benefits and challenges of modular integrated construction in Hong Kong: A literature review. In *The 8th international conference on construction engineering and Project Management, 7-8 Dec. 2020, Hong Kong SAR*.

Abdullah, O. M. A., Umut, H., & Al-Hadidy, A. I. (2021). Benefits of project management to realising sustainable buildings. *SSRN Electronic Journal*. https://doi.org/10.2139/ssrn.3873150

Abera, Y. A. (2024). Sustainable building materials: A comprehensive study on eco-friendly alternatives for construction. *Composites and Advanced Materials, 33*, 26349833241255957.

Abidin, N. Z., & Ijias, S. (2006). Involvement of sustainability issues in value management: Requisite factors. *ICCI, 13*, 2006.

Adames, M. (2023). *How is technology improving construction sustainability?* Retrieved August 12, 2024, from https://www.environmentenergyleader.com/stories/how-is-technology-improving-construction-sustainability,4032#:~:text=Technology%20Helps%20Achieve%20Sustainable%20Construction,successful%20and%20sustainable%20construction%20projects

Adegbembo, T. F., Bamisaye, O. P., & Aghimien, D. O. (2016). Assessment of lean construction practice in the Nigerian construction industry. In *Joint international conference (JIC) on 21st century human habitat: Issues, sustainability and development, 21–24 march 2016, Akure, Nigeria*.

Aghimien, D. O., Aigbavboa, C. O., Aghimien, L. M., Thwala, W. D., & Ndlovu, L. (2020). Making a case for 3D printing for housing delivery in South Africa. *International Journal of Housing Market and Analysis, 3*(4), 565–581. https://doi.org/10.1108/IJHMA-11-2019-0111

Aghimien, D. O., Aigbavboa, C. O., & Thwala, W. D. (2019). Microscoping the challenges of sustainable construction materials in developing countries. *Journal of Engineering Design and Technology, 17*(6), 1110–1128.

Aigbavboa, C. O., Ohiomah, I., & Zwane, T. (2017). Sustainable construction practices: "A lazy view" of construction professionals in the South African construction industry. *Energy Procedia, 105*, 3003–3010.

Aladeloba, E. A., Okesota, J. O., & Olawale, S. O. (2015). Prospects of industrialised materials in building production in Nigeria. *International Journal of Development Research, 5*(1), 3143–3148.

Al-Amin, F., & Taleb, H. (2016). Biomimicry approach to achieving thermal comfort in a hot climate. In: *Proceedings of sustainable built environment (SBE) 16 conference, Dubai, 17-19 January 2016* (pp. 1–8).

Al-Sanad, S. (2015). Awareness, drivers, actions, and barriers of sustainable construction in Kuwait. *Procedia Engineering, 118*, 969–983.

Altan, H., Gasperini, N., Moshaver, S., & Frattari, A. (2015). Redesigning terraced social housing in the UK for flexibility using building energy simulation with consideration of passive design. *Sustainability, 7*, 5488–5507. https://doi.org/10.3390/su7055488

Amaral, R. (2023). *Green certification. Global Sustainable Building Guide*. Retrieved August 2, 2024, from https://resourcehub.bakermckenzie.com/pl-pl/resources/global-sustainable-buildings/latin-america/brazil/topics/green-certification#:~:text=The%20accreditation%20is%20conducted%20by,to%20buy%20their%20own%20homes

Ametepey, O., & Aigbavboa, C. (2014). Practitioners' perspectives for the implementation of sustainable construction in Ghana. In: *Proceedings of the DII-2014 conference on infrastructure investments in Africa, Livingstone, Zambia, 25-26 September* (pp. 114–124).

Archistar. (2024). *How to adopt sustainable and greener building practices*. Retrieved August 12, 2024, from https://www.archistar.ai/blog/how-to-adopt-sustainable-and-greener-building-practices/#popup-id-31813

Asante, R., Faibil, D., Agyemang, M., & Ahmed Khan, S. (2022). Life cycle stage practices and strategies for circular economy: Assessment in the construction and demolition industry of an emerging economy. *Environmental Science and Pollution Research, 82*, 110–121.

Asif, M., Naeem, G., & Khalid, M. (2024). Digitalisation for sustainable buildings: Technologies, applications, potential, and challenges. *Journal of Cleaner Production, 450*, 1–19.

Ayoola, H. A., & Aghimien, E. I. (2017). Assessing the level of adoption of industrialised building system among professionals in Ondo state, Nigeria. In: *Proceedings of the environmental design and management international conference, held at Obafemi Awolowo University, Ile-Ife, Osun State, Nigeria, 22-24 May*.

Azazga, I. (2019). Economic benefits of sustainable construction. *Finance and Business Economics Review, 3*(1), 1–8.

Azeem, S., Naeem, M. A., & Waheed, A. (2020). Adoption of green building practices in Pakistan: Barriers and measures. In Z. Gou (Ed.), *Green building in developing countries. Green energy and technology*. Springer.

Azman, M. N. A., Ahmad, M. S. S., & Hussin, W. M. A. (2012). Comparative study on prefabrication construction process. *International Surveying Research Journal, 2*(1), 45–58.

Babuka, H. (2016). Application of nanotechnologies in sustainable architecture. In *Proceedings of fourth student conference "energy efficiency and sustainable development"*. SCEESD.

Bai, C., & Sarkis, J. (2020). A supply chain transparency and sustainability technology appraisal model for blockchain technology. *International Journal of Production Research, 58*(7), 2142–2162.

Bal, M., Bryde, D., Fearon, D., & Ochieng, E. (2013). Stakeholder engagement: Achieving sustainability in the construction sector. *Sustainability, 5*, 695–710.

Barbulianno. (2020). *18 eco-friendly building materials that help you save energy and the earth*. Retrieved August 2, 2024, from https://www.barbuliannodesign.com/post/eco-friendly-building-materials-list

Barnes, G. (2023). *Is digital construction sustainable?* Retrieved August 18, 2024, from https://www.planradar.com/gb/is-digital-construction-sustainable/

Baron, N., & Donath, D. (2016). Learning from Ethiopia—A discussion on sustainable building. In: *Proceedings of SBE16 Hamburg international conference on sustainable built environment strategies—Stakeholders—Success factors*, held from 7th to 11th march in Hamburg.

Bendi, D., Rana, M. Q., Arif, M., Goulding, J. S., & Sawhney, A. (2021). An off-site construction readiness maturity model for the Indian construction sector. *Construction Innovation, 21*(1), 123–142.

Bertram, N., Fuchs, S., Mischke, J., Palter, R., Strube, G. & Woetzel, J. (2019). *Modular construction: From projects to products*. McKinsey and Company Report. Retrieved August 12, 2024, from https://www.mckinsey.com/~/media/mckinsey/business%20functions/operations/our%20insights/modular%20construction%20from%20projects%20to%20products%20new/modular-construction-from-projects-to-products-full-report-new.pdf

Bibri, S. E. (2018). The IoT for smart sustainable cities of the future: An analytical framework for sensor-based big data applications for environmental sustainability. *Sustainable Cities and Society, 38*, 230–253.

BigRentz. (2023). *20 sustainable building materials for a greener future*. Retrieved August 12, 2024, from https://www.bigrentz.com/blog/sustainable-construction?srsltid=AfmBOop2wceZn6sLTl3hTYBVh_R39mZUd3YUp_jZkUQeWjRWIGQv_IYD

Biomimicry Institute. (2024). *What is biomimicry?* Retrieved August 12, 2024, from https://biomimicry.org/inspiration/what-is-biomimicry/#:~:text=For%20all%20the%20challenges%20we,with%20all%20life%20on%20earth

Bora, N., Doloi, H., & Crawford, R. (2023). Barriers to implementation of sustainable construction in India. In: *Proceedings of the 6th international conference on smart villages and rural development, the University of Melbourne, Victoria 3010, Australia, 4-5 Dec.*

Brundtland Report. (1987). *Report of the world commission on environment and development: Our common future*. United Nations General Assembly, Development and International Co-operation: Environment.

Build Chain. (2024). *How technology is improving the sustainability of the construction industry*. Retrieved August 18, 2024, from https://www.thebuildchain.co.uk/news/how-technology-is-improving-the-sustainability-of-the-construction-industry/

Capra, F., & Jakobsen, O. D. (2017). A conceptual framework for ecological economics based on systemic principles of life. *International Journal of Social Economics, 44*(6), 831–844.

Chen, X., Yang, H., & Lu, L. (2015). A comprehensive review on passive design approaches in green building rating tools. *Renewable and Sustainable Energy Reviews, 50*, 1425–1436.

Cheng, R.-D., Soebarto, V., Zhao, Z. Y., & Zillante, G. (2016). Facilitating the transition to sustainable construction: China's policies. *Journal of Cleaner Production, 131*, 534–544.

Cohen-Shacham, E., Walters, G., Janzen, C., & Maginnis, S. (2016). *Nature-based solutions to address global societal challenges*. IUCN.

Compliance Chain. (2024). *Sustainability in construction: Why it matters*. Retrieved December 1, 2024, from https://compliancechain.com/sustainability-in-construction-why-it-matters/#:~:text=The%20environmental%20benefits%20of%20sustainability,resources%2C%20and%20reduce%20energy%20consumption

Čuš-Babič, N., Rebolj, D., Nekrep-Perc, M., & Podbreznik, P. (2014). Supply-chain transparency within industrialised construction projects. *Computers in Industry, 65*(2), 345–353.

Da Silva, F. (2020). *Digitalisation is key for a net-zero carbon*. Retrieved August 12, 2024, from https://cmarkits.com/archives/6089

Dagou, H.H. & Gurgun, A.P. (2024). Blockchain and AI for sustainable supply chain management in construction. In: Kahraman, C., Cevik Onar, S., Cebi, S., Oztaysi, B., Tolga, A.C. & Ucal, I. (eds.) Intelligent and fuzzy systems. INFUS 2024. Lecture notes in networks and systems, vol. 1090. Springer, .

Davies, R. J., Pratama, M. M. A., & Yusuf, M. (2018). BIM adoption towards the sustainability of the construction industry in Indonesia. *MATEC Web of Conferences, 195*, 06003.

Djokoto, S. D., Dadzie, J., & Ohemeng-Ababio, E. (2014). Barriers to sustainable construction in the Ghanaian construction industry: Consultants' perspectives. *Journal of Sustainable Development, 7*(1), 134–143.

Dobias, J., & Macek, D. (2014). Leadership in energy and environmental design (LEED) and its impact on building operational expenditures. *Procedia Engineering, 85*, 132–139.

Du Plessis, C. (2002). *Agenda 21 for sustainable construction in developing countries - A discussion document*. International Council for Research and Innovation in Building and Construction, CIB. Retrieved August 12, 2024, from http://researchspace.csir.co.za/dspace/handle/10204/3511

Eggermont, H., Balian, E. V., Azevedo, J. M. N., Beumer, V., Brodin, T., Claudet, J., Fady, B., Grube, M., Keune, H., & Lamarque, P. (2018). Nature-based solutions: New influence for environmental management and research in Europe. *GAIA - Ecological Perspectives for Science and Society, 24*(4), 243–248.

Elrayies, G. M. (2018). Microalgae: Prospects for greener future buildings. *Renewable and Sustainable Energy Reviews, 81*, 1175–1191. https://doi.org/10.1016/j.rser.2017.08.032

Ershadi, M., & Goodarzi, F. (2021). Core capabilities for achieving sustainable construction project management. *Sustainable Production and Consumption, 28*, 1396–1410.

Eze, E. C., Aghimien, D. O., Aigbavboa, C. O., & Sofolahan, O. (2022). Building information modelling adoption for construction waste reduction in the construction industry of a developing country. *Engineering, Construction and Architectural Management, 31*(6), 2205–2223.

Eze, E. C., Sofolahan, O., & Omoboye, O. G. (2023). Assessment of barriers to the adoption of sustainable building materials (SBM) in the construction industry of a developing country. *Frontiers in Engineering and Built Environment, 3*(3), 153–166.

Farhadi, F. (2023). *Explore how nanotechnology in construction is revolutionising the industry, driving sustainability, and clearing the way for a resilient future*. Retrieved December 1, 2024, from https://neuroject.com/nanotechnology-in-construction/#:~:text=Environmental%20Impact%20and%20Sustainability,-Reduced%20Carbon%20Footprint&text=Nanotechnology%20in%20construction%20has%20the%20ability%20to%20significantly%20reduce%20the,frequency%20of%20repairs%20and%20replacements

Ferdous, W., Bai, Y., Ngo, T. D., Manalo, A., & Mendis, P. (2019). New advancements, challenges and opportunities of multi-storey modular buildings—A state-of-the-art review. *Engineering Structures, 183*, 883–893.

Fern, M. (2024). *Leveraging technology to drive sustainability in construction*. Retrieved December 1, 2024, from https://arkance.world/au-en/resources/read/sustainability/leveraging-technology-to-drive-sustainability-in-construction-world-zero-emissions-day#:~:text=Make%20Data%2DDriven%20Decisions:%20Utilize,decisions%20throughout%20the%20project%20lifecycle

Ford, S., & Despeisse, M. (2016). Additive manufacturing and sustainability: An exploratory study of the advantages and challenges. *Journal of Cleaner Production, 137*, 1573–1587.

Freitag, A. E. B., Quelhas, O. L. G., Franca, S. L. B., & Meirino, M. J. (2020). Implementation of lean management: Sustainability in the construction industry: The case of Brazilian companies. *Accounting, Finance, Sustainability, Governance and Fraud, 1*, 25–44.

Gehlot, M., & Shrivastava, S. (2022). Sustainable construction practices: A perspective view of Indian construction industry professionals. *Materials Today: Proceedings, 61*, 315–319.

Green, E. (2023). *Best sustainable materials for green buildings and constructions*. Retrieved August 12, 2024, from https://sigmaearth.com/best-sustainable-materials-for-green-buildings-and-constructions/#:~:text=Bamboo.%20Renowned%20for%20its%20rapid%20growth%20and,reducing%20carbon%20footprints%20and%20promoting%20resource%20conservation

Hafez, F. S., Sa'di, B., Safa-Gamal, M., Taufiq-Yap, Y. H., Alrifaey, M., Seyedmahmoudian, M., Stojcevski, A., Horan, B., & Mekhilef, S. (2023). Energy efficiency in sustainable buildings: A systematic review with taxonomy, challenges, motivations, methodological aspects, recommendations, and pathways for future research. *Energy Strategy Reviews, 45*, 1–30.

Hager, I., Golonka, A., & Putanowicz, R. (2016). 3D printing of buildings and building components as the future of sustainable construction? *Procedia Engineering, 151*, 292–299.

Halder, S., & Afsari, K. (2023). Robots in inspection and monitoring of buildings and infrastructure: A systematic review. *Applied Sciences, 13*, 2304.

Hallinger, P., & Kovačević, J. (2023). Applying bibliometric review methods in education: Rationale, definitions, analytical techniques, and illustrations. In *International encyclopedia of education* (pp. 546–556). Elsevier.

Hamid, Z. A., Kamar, K. A. M., Zain, M. Z. M., Ghani, M. K., & Rahim, A. H. A. (2008). Industrialised building system (IBS) in Malaysia: The current state and R&D initiatives. *MCRJ, CREAM, 2*(1), 1–11.

Hanafi, W. H. H. (2021). Bio-algae: A study of an interactive facade for commercial buildings in populated cities. *Journal of Engineering and Applied Science, 68*(37), 1–16.

International Energy Agency. (2019). *Global status report for buildings and construction 2019*. International Energy Agency. Retrieved August 12, 2024, from https://shorturl.at/dmzGZ

International Union for Conservation of Nature (IUCN). (2020). *The IUCN global standard for nature-based solutions* (1st ed.). IUCN. Retrieved August 12, 2024, from www.iucn.org/resources/publication/iucn-global-standard-nature-based-solutions-first-edition

Jin, R., Hong, J., & Zuo, J. (2020). Environmental performance of off-site constructed facilities: A critical review. *Energy and Buildings, 207*, 109567.

Joshi, S. C., & Sheikh, A. A. (2015). 3D printing in aerospace and its long-term sustainability. *Virtual and Physical Prototyping, 10*(4), 175–185.

Kaiv, T. (2023). *8 ways to create a more sustainable construction site*. Retrieved August 12, 2024, from https://www.bauwise.com/8-ways-to-create-a-more-sustainable-construction-site/#:~:text=Sustainable%20building%20materials%20are%20environmentally%20responsible%20throughout,construction%20by%20reducing%20carbon%20emissions%20and%20waste

Kamali, M., & Hewage, K. (2016). Life cycle performance of modular buildings: A critical review. *Renewable and Sustainable Energy Reviews, 62*, 1171–1183.

Karthikeyan, V., Vinodhini, E., Aparna, P., & Monika & Kumar, R.S. (2018). Study on comparison between prefabricated and conventional structures. *International Journal of Civil Engineering and Technology, 9*(5), 1–8.

Kazancoglu, I., Sagnak, M., Kumar Mangla, S., & Kazancoglu, Y. (2020). *Circular economy and the policy: A framework for improving corporate environmental management in supply chains* (pp. 590–608). Wiley Online Library.

Kern, F., Rogge, K. S., & Howlett, M. (2019). Policy mixes for sustainability transitions: New approaches and insights through bridging innovation and policy studies. *Research Policy, 48*, 103832.

Khalil, F. D. A. A., Aziz, F. N. A. A., Hassim, S., & Jaafar, M. S. (2016). A review on industrialised building system issues in Malaysia. *MATEC Web of Conferences, 47*, 04019.

Kibert, C. J. (2013). *Sustainable construction: Green building design and delivery*. John Wiley & Sons Inc.

Kiy, M. (2022). *Siemens*. Retrieved August 12, 2024, from https://www.siemens.com/global/en/company/stories/infrastructure/2022/7-major-trends-that-will-transform-the-construction-industry.html

Lambert, R. (2022). *HSE people*. Retrieved December 1, 2024, from https://www.hsepeople.com/sustainability-in-construction-how-ai-can-promote-a-multi-tier-sustainable-ecosystem/

Level Construction Recruitment. (2024). *The impact of sustainable construction practices on American cities*. Retrieved December 1, 2024, from https://lcr.careers/the-impact-of-sustainable-construction-practices-on-american-cities/#:~:text=One%20of%20the%20key%20indicators,and%20the%20WELL%20Building%20Standard

Li, C. Z., Xue, F., Li, X., Hong, J., & Shen, G. Q. (2018). An internet of things-enabled BIM platform for on-site assembly services in prefabricated construction. *Automation in Construction, 89*, 146–161.

Li, J., Greenwood, D., & Kassem, M. (2019). Blockchain in the built environment and CI: A systematic review, conceptual models and practical use cases. *Automation in Construction, 102*, 288–307.

Long, W., Bao, Z., Chen, K., Ng, S. T., & Wuni, I. Y. (2024). Developing an integrative framework for digital twin applications in the building construction industry: A systematic literature review. *Advances in Engineering Informatics, 59*, 102346.

Lu, Q., Xie, X., Parlikad, A. K., Schooling, J. M., & Konstantinou, E. (2020). Moving from building information models to digital twins for operation and maintenance. *Proceedings of the Institution of Civil Engineers: Smart Infrastructure and Construction, 174*, 46–56.

Lu, W., Lou, J., Ababio, B. K., Zhong, R. Y., Boa, Z., Li, X., & Xue, F. (2024). Digital technologies for construction sustainability: Status quo, challenges, and future prospects. *Material Sustainability, 2*(10), 1–10.

Lu, Y., Wu, Z., Chang, R., & Li, Y. (2017). Building information modelling (BIM) for green buildings: A critical review and future directions. *Automation in Construction, 83*, 134–148.

Ma, X., Yuan, H., & Du, W. (2024). Blockchain-enabled construction and demolition waste management: Advancing information management for enhanced sustainability and efficiency. *Sustainability, 16*(2), 721.

Mamillapalli, R. S., Rath, A. K., & Bera, D. K. (2021). Studies on integration of lean construction and sustainability in Indian real estate projects. *Lecture Notes in Civil Engineering, 75*, 951–959.

Masyhur, R. T., Alias, A. H., Haron, N. A., & Zulkafli, Z. (2024). A systematic review on green practices in the Malaysian construction industry: Status, challenges, key motivations, rating systems and technology evolution. *Energy and Buildings, 320*, 1–22. https://doi.org/10.1016/j.enbuild.2024.114550

McBride Sustainability. (2023). *Sustainability matters*. Retrieved August 12, 2024, from https://www.linkedin.com/pulse/circular-economy-benefits-strategies-implementing/

Musa, M. F., Mohammad, M. F., Mahbub, R., & Yusof, M. R. (2014). Enhancing the quality of life by adopting sustainable modular industrialised building system (IBS) in the Malaysian construction industry. *Procedia - Social and Behavioral Sciences, 153*, 79–89.

Nahmens, I., & Ikuma, L. H. (2011). Effects of lean construction on sustainability of modular homebuilding. *Journal of Architectural Engineering, 18*(2), 155–163.

Nawi, M. N. M., Pozin, M. A. A., Kamar, K. A. M., Lee, A., & Harun, A. N. (2019). The global adoption of industrialised building system (IBS): Lessons learned. *The Journal of Social Sciences Research, 6*, 1272–1278.

Neo, H. (2009). Resource and environmental economics. In *International encyclopedia of human geography* (pp. 376–380). Elsevier.

Ngcobo, N., Aigbavboa, C. O., & Aghimien, D. O. (2024). *Maintenance engineering for buildings: Digital solutions and sustainable practices*. Woodhead Publishing.

Nguyen, H., Skitmore, M., Gray, M., Zhang, M., & Olanipekun, A. O. (2017). Will green building development take off? An exploratory study of barriers to green building in Vietnam. *Resources, Conservation and Recycling, 127*, 82.

Noor, N. F., Kamruzzaman, S. N., & Ghaffar, N. (2015). Sustainability concern in value management: A study on government building projects. *International Journal of Current Research and Academic Review, 2*, 72–83.

Nwokediegwu, Z. Q. S., Ilojianya, V. I., Ibekwe, K. I., Adefemi, A., Etukudoh, E. A., & Umoh, A. A. (2024). Advanced materials for sustainable construction: A review of innovations and environmental benefits. *Engineering Science & Technology Journal, 5*, 744.

Oguntona, O. A., & Aigbavboa, C. O. (2017). Biomimetic reinvention of the CI: Energy management and sustainability. *Energy Procedia, 142*, 2721–2727.

Oke, A., Aigbavboa, C., & Ndou, M. (2018). Promoting sustainable construction through ecological economics. In: *FIG congress 2018, Istanbul, 6–11 May 2018* (pp. 1–10).

Oke, A. E., Aghimien, D. O., Aigbavboa, C. O., & Musenga, C. (2019). Drivers of sustainable construction practices in the Zambian construction industry. *Energy Procedia, 158*, 3246–3252.

Oke, A. E., Aghimien, D. O., & Olatunji, S. O. (2015). Implementation of value management as an economic sustainability tool for building construction in Nigeria. *International Journal for Managing Value and Supply Chain, 6*(4), 55–64.

Oke, A. E., Aigbavboa, C. O., & Semenya, K. (2017). Energy savings and sustainable construction: Examining the advantages of nanotechnology. *Energy Procedia, 142*, 3839–3843.

Olawumi, T. O., & Chan, D. W. (2018). Identifying and prioritising the benefits of integrating BIM and sustainability practices in construction projects: A Delphi survey of international experts. *Sustainable Cities and Society, 40*, 16–27.

Palomo, I., Locatelli, B., Otero, I., Colloff, M., Crouzat, E., Cuni-Sanchez, A., Gómez-Baggethun, E., González-García, A., Grêt-Regamey, A., Jiménez-Aceituno, A., Martín-López, B., Pascual, U., Zafra-Calvo, N., Bruley, E., Fischborn, M., Metz, R., & Lavorel, S. (2021). Assessing nature-based solutions for transformative change. *One Earth, 4*, 730–741.

Papadakis, N., & Katsaprakakis, D. A. (2023). A review of energy efficiency interventions in public buildings. *Energies, 16*(17), 6329.

Patel, P., & Patel, A. (2021). Use of sustainable green materials in construction of green buildings for sustainable development. *IOP Conference Series: Earth and Environmental Science, 785*, 012009.

Rockfon. (2023). *The best-known global green building certifications*. Retrieved February 4, 2023, from https://www.rockfon.co.uk/about-us/blog/2023/green-building-certifications/#:~:text=Platinum-,BREEAM%20Green%20Building%20Certification,BREEAM%20operators%20are%20not%20available

Saad, M. M. (2016). Impediments to implementation of green buildings in South Africa. In: *9th CIDB postgraduate conference, Cape Town, 2–4 February*.

Saberi, S., Sarkis, J., & Shen, L. (2020). Blockchain technology and its relationships to sustainable supply chain management. *International Journal of Production Research, 57*, 2117–2135.

Sacks, R., Girolami, M., & Brilakis, I. (2020). Building information modelling, artificial intelligence and construction tech. *Developments in the Built Environment, 4*, 100011.

Sajjad, M., Hu, A., Waqar, A., Falqi, I. I., Alsulamy, S. H., Bageis, A. S., & Alshehri, A. M. (2023). Evaluation of the success of industry 4.0 digitalisation practices for sustainable construction management: Chinese construction industry. *Buildings, 13*, 1668.

Sakin, M., & Kiroglu, Y. C. (2017). 3D printing of buildings: Construction of the sustainable houses of the future by BIM. *Energy Procedia, 134*, 702–711.

Shi, X., Tian, Z., Chen, W., Si, B., & Jin, X. (2016). A review on building energy efficient design optimisation from the perspective of architects. *Renewable and Sustainable Energy Reviews, 65*, 872–884.

Shojaei, A., & Naderi, H. (2024). Blockchain technology for a circular built environment. In C. De Wolf, S. Çetin, & N. M. P. Bocken (Eds.), *A circular built environment in the digital age. Circular economy and sustainability*. Springer.

Shuja, U. (2024). *Is digitalisation the key to sustainable construction?* Retrieved December 7, 2024, from https://www.forbes.com/councils/forbestechcouncil/2024/05/16/is-digitalization-the-key-to-sustainable-construction/#:~:text=Digitalization%20Is%20Part%20Of%20The,can%20reduce%20its%20environmental%20impact

Stanca, S. (2023). Sustainability in construction. *Metallurgy and Materials Science, 46*(1), 10–15.

Subramanya, K., Kermanshachi, S., & Rouhanizadeh, B. (2020). Modular construction vs. traditional construction: Advantages and limitations: A comparative study. *Proceedings of the Creative Construction E-conference*. https://doi.org/10.3311/CCC2020-012

Szkop, Z., Szewczyk, M., & Mikołajczyk, P. (2021). *Nature-based solutions for urban challenges*. UN Foresight Brief. Retrieved August 12, 2024, from https://unepdhi.org/wp-content/uploads/sites/2/2023/02/NbS_for_Urban_Challenges_UNEP_23rd_Foresight_Brief.pdf

Tan, L. (2024). *Exploring the impact of AI on sustainable building design and construction*. Retrieved August 12, 2024, from https://parametric-architecture.com/ai-impact-on-sustainable-building-design/#:~:text=Sustainable%20construction%20methods,-Chinese%20construction%20firms&text=AI%2Ddriven%20systems%20are%20revolutionizing,and%20glass%20are%20used%20efficiently

Tan, Y., Shen, L., & Yao, H. (2011). Sustainable construction practice and contractors' competitiveness: A preliminary study. *Habitat International, 35*(2), 225–230.

Tao, X., Mao, C., Xie, F., Liu, G., & Xu, P. (2018). Greenhouse gas emission monitoring system for manufacturing prefabricated components. *Automation in Construction, 93*, 361–374.

Teh, D., Tehmina, K., Corbitt, B., & Ong, C. E. (2020). Sustainability strategy and blockchain-enabled life cycle assessment: A focus on materials industry. *Environment Systems and Decisions*, 605–622.

Thounaojam, N., Devkar, G., & Laishram, B. (2023). Sustainability practices implemented in the Indian construction industry: A focus of construction phase. In Y. G. Sandanayake, K. G. A. S. Waidyasekara, T. Ramachandra, & K. A. T. O. Ranadewa (Eds.), *Proceedings of the 11th world construction symposium* (pp. 1011–1023). Ceylon Institute of Builders.

United Nations Environment Assembly (2020). *Resolution adopted by the United Nations Environment Assembly on 2 March 2022 — Nature-based solutions for supporting sustainable development, UNEP/EA.5/Res.5*. Retrieved August 12, 2024, from https://wedocs.unep.org/bitstream/handle/20.500.11822/39752/K2200677%20-%20UNEP-EA.5-Res.5%20-%20Advance.pdf?sequence=1&isAllowed=y

United Nations Environment Programme (2020). *2020 global status Report for buildings and construction: Towards a zero-emission, efficient and resilient buildings and construction sector*. Global Status Report. United Nations. Retrieved December 1, 2024, from www.iea.org

Vasilca, I.-S., Nen, M., Chivu, O., Radu, V., Simion, C.-P., & Marinescu, N. (2021). The management of environmental resources in the construction sector: An empirical model. *Energies, 14*(9), 1–19.

Wang, M., Wang, C. C., Sepasgozar, S., & Zlantanova, S. (2020). A systematic review of digital technology adoption in off-site construction: Current status and future direction towards industry 4.0. *Buildings, 10*, 204.

Wasim, M., Han, T. M., Huang, H., Madiyev, M., & Ngo, T. D. (2020). An approach for sustainable, cost-effective, and optimised material design for the prefabricated non-structural components of residential buildings. *Journal of Building Engineering, 32*, 101474.

Weniger, A., Del Rosario, P., Backes, J. G., & Traverso, M. (2023). Consumer behavior and sustainability in the construction industry—Relevance of sustainability-related criteria in purchasing decisions. *Buildings, 13*(3), 638.

Wickenberg, B., McCormick, K., & Olsson, J. A. (2021). Advancing the implementation of nature-based solutions in cities: A review of frameworks. *Environmental Science & Policy, 125*, 44–53.

Wijewansha, A. S., Tennakoon, G. A., Waidyasekara, K. G. A. S., & Ekanayake, B. J. (2021). Implementation of circular economy principles during preconstruction stage: The case of Sri Lanka. *Built Environment Project and Asset Management, 11*(4), 750–766.

Wolf, C. D. (2022). *4 promising digital technologies for circular construction*. Retrieved December 1, 2024, from https://www.weforum.org/agenda/2022/09/4-promising-digital-technologies-for-circular-construction/?DAG=3&gclid=Cj0KCQjw1_SkBhDwARIsANbGpFtBwqUzmAl7dLj7oqeM1tOYDyGzCNgLxT6a0vFkZlPmA4qvB-QwZAgaAj3cEALw_wcB

Wood, C. (2024). *Top 25 sustainable building materials*. Retrieved August 12, 2024, from https://www.boomandbucket.com/blog/sustainable-building-materials

World Economic Forum. (2022a). *Digital solutions can reduce global emissions by up to 20%. Here's how*. Retrieved August 12, 2024, from https://www.weforum.org/agenda/2022/05/how-digital-solutions-can-reduce-global-emissions/

World Economic Forum. (2022b). *3D printing can help the construction industry become carbon-neutral. Here's how*. Retrieved August 12, 2024, from https://www.weforum.org/stories/2022/06/3d-printing-can-help-make-construction-carbon-neutral-heres-how/#:~:text=3D%20printing%20is%20an%20industry,that%203D%20printing%20can%20deliver

World Green Building Council. (2019). *Bringing embodied carbon upfront*. Retrieved August 12, 2024, from https://worldgbc.org/advancing-net-zero/embodied-carbon/

Wu, Y., Wu, Y., Guerrero, J. M., Vasquez, J. C., Palacios-García, E. J., & Guan, Y. (2020). IoT-enabled microgrid for intelligent energy-aware buildings: A novel hierarchical self-consumption scheme with renewables. *Electronics, 9*(4), 550. https://doi.org/10.3390/electronics9040550

Wuni, I. Y., & Shen, G. Q. (2020). Key success factors for implementing modular integrated construction projects: A literature mining approach. In: *The 8th international conference on construction engineering and Project Management, 7–8 Dec. 2020, Hong Kong SAR*.

Xero. (2024). *What is sustainable construction?* Retrieved December 1, 2024, from https://www.xero.com/uk/guides/what-is-sustainable-construction/

Xu, J., Lou, J., Lu, W., Wu, L., & Chen, C. (2023). Ensuring construction material provenance using internet of things and blockchain: Learning from the food industry. *Journal of Industrial Information Integration, 33*, 100455.

Xu, Z., Zayed, T., & Niu, Y. (2019). Comparative analysis of modular construction practices in mainland China, Hong Kong and Singapore. *Journal of Cleaner Production, 245*, 118861.

Yaman, Y., Tokuç, A., Sener, I., Altunacar, N., Köktürk, G., Deniz, I., & Ezan, M. A. (2021). Energy efficient buildings with algae. *IOP Conference Series: Earth and Environmental Science, 798*(1), 1–9. https://doi.org/10.1088/1755-1315/798/1/012001

Yang, Z., Chen, H., Mi, L., Li, P., & Qi, K. (2021). Green building technologies adoption process in China: How environmental policies are reshaping the decision-making among alliance-based construction enterprises? *Sustainable Cities and Society, 73*, 1–15. https://doi.org/10.1016/j.scs.2021.103122

Yang, Z., Zhu, C., Zhu, Y., & Li, X. (2023). Blockchain technology in building environmental sustainability: A systematic literature review and future perspectives. *Building and Environment, 245*, 110970.

Yekinni, A. A., Bello, S. K., & Olaiya, K. A. (2015). Application of value engineering techniques in sustainable product and service design. *Science and Engineering Perspectives, 10*, 120–130.

Chapter 5
Ethical Practices in Digitalisation

Introduction

Due to the quick uptake of digital technology, the construction industry is poised for significant change. Technologies, such as Radio Frequency Identification (RFID) (Lu et al., 2011), robotics (Aghimien et al., 2020), Blockchain technology (Turk & Klinc, 2017), Building Information Modelling (BIM) (Sacks et al., 2018), Internet of Things (IoT) (Patil, 2019), and sensors (Akinci & Anumba, 2008), are just a few examples of how digitalisation is changing the way infrastructure and buildings are planned, built, and operated. This digital revolution promises unprecedented advancements in efficiency, safety, and sustainability. Integrating digital technologies in construction offers a compelling vision for the future. Projects can be executed more swiftly, with greater precision and cost-effectiveness. Building Information Modelling (BIM) facilitates a collaborative approach to design and construction, enabling stakeholders to visualise and modify projects in real time. Artificial Intelligence (AI) algorithms can optimise project schedules, predict and mitigate risks, and enhance overall project management. IoT-sensors continuously monitor building sites and structures, ensuring safety and performance.

However, despite the benefits of digitalisation, it also presents difficult moral dilemmas that need to be resolved to guarantee that the advantages of new technology are distributed fairly and responsibly. Data privacy and security are critical concerns, given the extensive data collection required for smart buildings and advanced project management systems. The potential for data breaches or misuse of information calls for stringent security measures and clear policies on data ownership and access. Modern technology and automation have the potential to replace traditional jobs while also opening new career opportunities that call for various skill sets. This change creates two challenges: making sure that the advantages of

O. Oladinrin et al., *Ethical Practices for Sustainable Construction Digitalisation*, https://doi.org/10.1007/978-3-032-00151-1_5

digitalisation are shared fairly and avoiding the marginalisation of workers who might not be tech-savvy. Therefore, as we embrace digitalisation, we must confront crucial ethical questions: How can we safeguard the enormous volumes of data created and utilised in the construction process? What policies can we implement to guarantee accountability and openness in technology-driven decision-making? How do we manage the socio-economic impacts on the workforce, particularly regarding job displacement and the need for new skills? These are not just theoretical concerns; they are fundamental to the continuous change in the construction industry and call for prompt, careful action.

The Concepts of Ethics

According to McCarthy (2012), understanding the definition of ethics and its historical progression is crucial for any ethical assessment. 'Ethics' and 'morality' are frequently regarded as interchangeable. However, they can sometimes be differentiated in that morality pertains to an intricate framework of rules, values, and norms that dictate or are intended to govern individuals' behaviour, while ethics involves the theoretical examination of morality underpinned by moral philosophy. According to Fisher and Lovell (2009), ethics is a field of philosophy that deals with formal academic reasoning about what is right and wrong. McDonald (2004) posited that ethics is analysing whether something is good or bad. Researchers such as Boatright (2007), Crane and Matten (2010), and Thompson (2005) believe that ethics refers to right and wrong acts and the consequent reactions. Ethics is the study of morals and the norms that society acknowledges as wrong or right and good or bad (De George, 2006).

In the construction industry context, Oladinrin (2016) defined ethics as moral standards that can influence the behaviours of employees and other stakeholders within construction organisations. This definition was given from the perspective of 'ethics codes' as a tool, rules, principles, philosophy, and conduct set down to guide the behaviours of individuals in an organisation or society. Wempe and Kaptein (2002) described ethical codes as policies that outline an organisation's duties to stakeholders and the anticipated behaviour of its members. These documents outline the organisation's ethical guidelines and define acceptable and inappropriate behaviours. The literature on ethics encompasses three primary branches of philosophical study: meta-ethics, normative ethics, and applied ethics.

Meta-ethics

Meta-ethics focuses on the reasoning behind the justification of moral judgements rather than on forming those judgements themselves (DeLapp, 2011). This means it elucidates the meaning of moral statements and essentially explores the nature of morality. Meta-ethics specifically captures issues related to existence (ontology), meaning (semantics), and knowledge (epistemology). Moral ontology examines the aspects of the world that hold moral relevance or value. Moral semantics explores the definitions of moral concepts like right, wrong, good, bad, and ought, which are among the most significant. Moral epistemology investigates how we can ascertain moral truths.

Normative Ethics

Normative ethics is concerned with developing philosophical theories that evaluate actions to determine what is right and wrong. Various ethical theories, such as deontological, utilitarian, and virtue ethics, are examples of normative ethics (Hursthouse, 2013).

Deontological concepts addressed morality as based on a specific set of duties and responsibilities, where the ethical value of an action is determined by the principles or norms that guide the agent's behaviour rather than the impact of those actions on others or their consequences for human well-being (Waller, 2011). Numerous ethical guidelines concerning digital technologies like AI and robotics can be attributed to a deontological perspective on ethics, which establishes universal principles that direct the ethical development, implementation, and governance of these tools (Eurofound, 2023).

Utilitarianism, a leading form of consequentialism, supports making ethical choices that enhance the collective good instead of prioritising individual interests. An outcome is viewed as favourable if it promotes overall welfare more effectively than any other available option (Shaw, 2017). Nevertheless, utilitarianism encounters both a measurement challenge (how to assess utility) and a conceptual issue (how to define utility for society) when applied to the ethical problems posed by powerful technologies like AI.

Virtue ethics can be traced back to Greek philosophers, particularly Plato, who introduced the cardinal virtues of wisdom, justice, fortitude, and temperance. Aristotle further elaborated on this by outlining 11 moral virtues and incorporating intellectual virtues as well. The virtue ethics perspective emphasises striving for goodness and highlights the significance of guiding individuals to cultivate a virtuous character that serves as a benchmark for ethical behaviour and decision-making. This viewpoint has shaped ethical deliberations in the realm of information technology (Stahl, 2021) and has established the groundwork for what is referred to as 'flourishing ethics' (Bynum, 2006). It prioritises human values and principles, such as freedom, autonomy, equality, justice, and benevolence—that support human flourishing in a technology-driven world. The core notion is that technologies should be created and utilised to enhance human well-being or benefit society (Powell et al., 2022).

Applied Ethics

Applied ethics, known as practical ethics, involves using ethical principles to address real-world issues. Practical ethics seeks to determine how individuals ought to behave in particular circumstances. For instance, is it morally acceptable for a construction firm to bluff during contract negotiations with the client? Applied ethics pertains to specific areas where ethical decisions are rendered by applying normative ethical theories. Petersen and Ryberg (2019) described applied ethics as a field of ethics that focuses on addressing moral challenges, practices, and policies in various aspects of personal life, professions, technology, and government. As ethical frameworks are being developed to tackle moral concerns related to digitalisation, digital ethics aligns closely with applied ethics.

Ethical Practices in Construction

Some researchers have investigated ethics management in the construction industry. Vee and Skitmore (2003) revealed that there is a growing need for ethical practices in the construction business and recommended that proper implementation and enforcement of ethical norms are necessary for growth. Poon (2004) emphasised the necessity of ethics in the construction business due to its unique nature and characteristics. Oladinrin and Ho (2016) proposed fully integrating codes into construction organisations' regular actions to foster ethical behaviours. Suen et al. (2007) revealed that profit-making, maximising economic prospects, and practising complete entrepreneurship are the primary objectives of company managers, which are pursued at the expense of ethical conduct. This remark reflects a prevalent trend in construction organisations characterised by intense pressure to make quick profits (Oladinrin & Ho, 2016). The fact that company owners are less concerned with ethics does not negate the intrinsic repercussions of being unethical. Suen et al. (2007), in their research on managing ethical conduct in construction organisations, posited that ethics is necessary because it enables individuals to appropriately place themselves within the web of interrelationships when confronted with real-world challenges. Construction operations impact everyone. Thus, disregarding any misconduct of the participants is risky in construction practices (Oladinrin & Ho, 2016).

Ethical Practices and Digitalisation

Recent innovations and technological advancements are transforming the construction industry towards automation, optimisation, and digitalisation. The transformation is bringing about profound changes in how people live and work. However, the possible harms, prejudices, and discriminations inherent in and created by such technologies in ethical and social contexts are not adequately considered (Khodabakhshian, 2024). Depending on their usage and purpose, digital technologies can lead to both societal benefits and growth, but they can also raise ethical concerns. The ongoing digitisation of society has highlighted the importance of ethical issues in the creation and application of technology. The issues are intrinsically related to the consequences of digitalising processes and automating jobs to improve working conditions. The widespread integration of digital technologies in today's society, manifested in their significance in nearly every aspect of daily life for average individuals, has led to major concerns regarding privacy, security, safety, sustainability, justice, inequality, and discrimination.

The workplace is an essential arena in which transformational potential of technologies may be demonstrated, yet it is sometimes overlooked in policy discussions about ethics and digital technology (Eurofund, 2023). In workplaces that are driven by data, interconnected digitally, and automated, fresh ethical dilemmas emerge. These concerns include, for instance, the collection and handling of personal data, the evolving role of human workers, and the feelings of distress or anxiety among employees who view their jobs as potentially threatened. Simultaneously, the line between the physical world and the individual worker is becoming progressively blurred, resulting in a series of ethical issues and several obstructions to institutional system expansion. There are two realms of advancement in technology (Chiodo et al., 2024). The first realm focuses on solving problems by principle using algorithms developed from science, technology, engineering, and mathematics (STEM) knowledge. The second realm addresses other aspects of life, such as social problems, value issues, and philosophical or religious matters. It is common in practice to see engineers who develop algorithms completely disinterested in the second realm by either disregarding social and ethical matters or delegating the issues to other specialists (Chiodo et al., 2024). The ethical application of technology involves not just following ethical standards and respecting fundamental rights but also foreseeing and alleviating adverse effects on working conditions. Assessing the risks related to technology implementation on working conditions, human rights, and ethical standards should be conducted thoroughly as an initial step in the design and development process. Ethics play a crucial role in how technology is embraced and utilised in professional settings and is essential to the overall quality of work.

Understanding Ethical Practices and Digitalisation from the Perspective of Past Studies

The relationship between digitalisation and ethics was further explored through the review of past studies. A bibliometric approach is a systematic analysis of literature published within a selected area using quantitative tools and map creation (Hallinger & Kovačević, 2023). For a bibliometric review to be conducted, bibliographic data were first gathered from the Scopus database. Careful selection of appropriate keywords was done, and these were searched in the database using two Boolean operators ('OR', 'AND'). The search protocol followed: TITLE-ABS-KEY ('ethics' OR 'ethics of technology' OR 'Business ethics' OR 'Digital ethics' AND 'digitalisation' OR 'Digital technologies' OR 'Digital transformation'. The search was limited to 2014–2024 to capture documents published in the last 10 years in order to gain recency. The initial search revealed 1622 documents. To ensure only relevant papers were extracted, the result was further refined in the subject area (engineering, computing, social science, and business management) to get a broader view of these concepts within construction-related domains. The language of publication was set at English. Based on these new parameters and screening for duplication or incomplete details, 987 documents emerged and were used for the bibliometric analysis.

The overall information of the bibliographic data revealed that the 987 documents contain 487 articles, 187 conference papers, 93 books, 140 book chapters, 12 editorials, and 68 reviews. These documents have a total of 2639 authors, with 274 single-authored documents. Figure 5.1 shows the frequency of publication. The figures show a steady increase in the publication per year from 2014 to 2023, with a single document drop in 2018. The year 2024 witnessed 189 documents published, and there is a possibility of more published documents before the end of the year, as the data used for the analysis was extracted in October 2024. This steady rise in documents over the years implies an increased interest in this topic and its importance.

Further observation of the countries of publication revealed that the extracted documents came from 87 different countries, with the UK having the highest number of publications of 363, with total citations of 3165 and an average article citation of 41.1. Next is the USA (308 documents; 833 citations), Germany (253 documents; 664 citations), Australia (138 documents; 319 citations), India (98 documents; 93 citations), Switzerland (94 documents; 152 citations), and Italy (90 documents; 88 citations). This implies that this area of research has received attention from diverse parts of the world, a further testament to its importance.

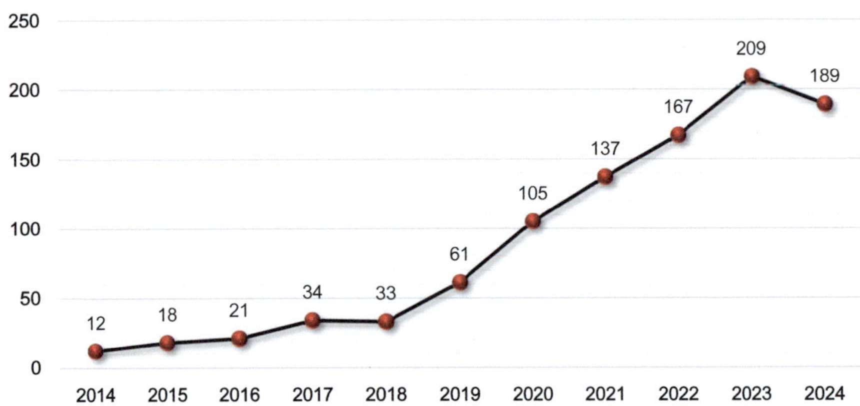

Fig. 5.1 Publication per year of ethics and digitalisation studies

In identifying the most impactful documents in this area of research, the citations gathered for each document were analysed. The result revealed that the work of Dwivedi et al. (2023), which explored a multidisciplinary perspective of the opportunities, challenges, and implications of generative conversational AI, is the most influential, with 1409 citations. The article was published in the International Journal of Information Management. This is followed by the work of Floridi and Taddeo (2016) on what is digital ethics. This article has garnered 274 citations since being published. Also, Saarikko et al. (2020) provided five crucial recommendations for digitally conscious firms, including taking responsibility for data ownership and ethics. This article, published in the Journal of Business Horizons, has been cited 265 times. Ashok et al. (2022) proposed an ethical framework for AI and digital technologies by reviewing extant literature. The article was published by the International Journal of Information Management and has been cited 170 times already. Similarly, the works of Bryson and Winfield (2017), Burr et al. (2020), and Wach et al. (2023) have all gained significant citations of 154, 145, and 137, respectively. Bryson and Winfield (2017) strived to standardise the ethical design of AI and autonomous systems to effectively guide the way technology impacts society. Burr et al. (2020), on the other hand, employed a thematic review to determine the ethics of digital well-being. Wach et al. (2023) further assessed the dark side of generative AI by exploring the controversies and risks of ChatGPT.

To determine the major areas of focus of these extracted documents, a word cloud of the top 100 terms was analysed using Biblioshiny in R-Studio and presented in Fig. 5.2. The analysis revealed that common terms were digital technologies (f = 104), human (f = 104), ethics (f = 98), AI (f = 84), ethical technology (f = 61), digital transformation (f = 46), and privacy (f = 32).

Further analysis of common terms was done using the visualisation of similarities viewer (VOSviewer) software. This software allows a network visualisation of the most co-occurring terms to be conducted. Setting the minimum co-occurrence

Fig. 5.2 Word cloud of common terms in ethics and digitalisation studies

at 10, a total of 94 terms clustered into four groups, as seen in Fig. 5.3. These four clusters had 1923 links to one another and 5629 total link strength. The four clusters are presented below.

Cluster 1: Ethical concerns in digital technology utilisation. The first cluster in Region 1 on the map has 36 co-occurring terms. The major terms in this cluster are AI (147 occurrences, 91 links), digital technologies (144 occurrences, 86 links), philosophical aspects (82 occurrences, 72 links), and digital transformation (78 occurrences, 72 links). Other prominent terms are ethical technology, digital ethics, ethical concerns, data privacy, metadata, IoT, risk management, security, cybersecurity, industry 4.0, and machine learning, among others. Based on the similarity of these terms, this cluster was named 'ethical issues in digital technology utilisation'.

Cluster 2: Ethical implication of digital technology integration in the health and well-being of humans. The second cluster in Region 2 on the map revealed 32 co-occurring terms. Prominent terms in this cluster are human (104 occurrences, 74 links), digital technology (90 occurrences, 75 links), and privacy (42 occurrences, 72 links). Other key terms in this cluster are mental health, health care delivery, medical ethics, telehealth, health care system, telemedicine, public health, bioethics, research ethics, human experiment, information technology, responsibility, social justice, education, and digital literacy, among others. These key terms relate to studies exploring the 'ethical implication of digital technology integration in health and well-being of humans'.

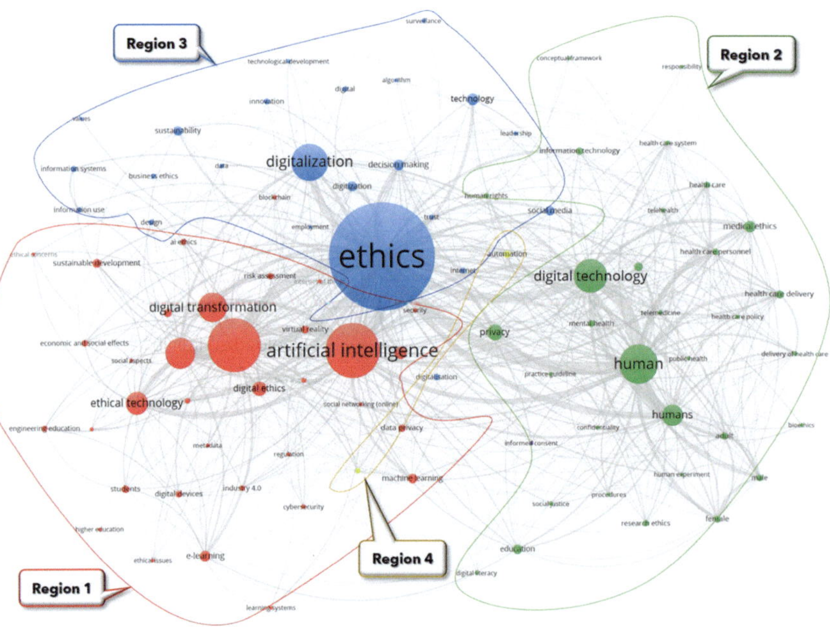

Fig. 5.3 Network visualisation of co-occurring terms in ethics and digitalisation studies

Cluster 3: Ethical use of digitalisation in organisation's decision-making. The third cluster in Region 3 on the map revealed 23 co-occurring terms. Ethics (295 occurrences, 93 links) and digitalisation (100 occurrences, 73 links) are at the core of the cluster. The term ethics also shows a strong link between terms in other clusters like AI (link strength of 55) human (link strength of 59) and digital technologies (link strength of 54). Other key terms in this cluster include business ethics, data, decision-making, design, information systems and use, internet, innovation, social media, surveillance, technology, trust, and values, among others. This cluster was subsequently named 'ethical use of digitalisation in organisation's decision making'.

Cluster 4: Ethical integration of automation and human-computer interaction. The last cluster in Region 4 on the map only two co-occurring terms, which are automation (19 occurrences, 43 links), human-computer interaction (15 occurrences, 4 links), and informed consent (11 occurrences, 5 links). This cluster refers to the 'ethical integration of automation and human-computer interaction'.

These four clusters revealed the significant attention garnered by ethics and digitalisation over the last 10 years. Digital technologies, including AI, machine learning, and IoT, have been evaluated for their ethical usage and influence on decision-making. The ethical use of emerging digital technologies on the health and well-being of people, as well as the ethical human-computer interactions, have been explored.

Factors Influencing Ethics in Digitalisation

Construction firms are making significant strides as the sector recognises the necessity to evolve and adapt its conventional business model. Driven by the technological advancements of the fourth industrial revolution, also called Industry 4.0 or its construction equivalent, Construction 4.0, the imperative to explore and effectively utilise technologies has become increasingly essential for those involved in construction research and practice (Hatoum & Nassereddine, 2024). The integration and evolution of digital technologies like AI, BIM, the IoT, blockchain, big data, and cloud models in the construction industry have addressed numerous issues and barriers faced in traditional construction methods (Hosseini et al., 2020). These advancements provide essential technologies and foundational platforms that enable construction companies to pursue digital transformation and achieve sustainable growth (Li et al., 2022).

The distinct features of the construction industry, especially its high level of fragmentation, create obstacles to the adoption of these technologies and achieving their maximum potential (Muñoz-La Rivera et al., 2021). The industry exhibits vertical fragmentation among various trade specialities, horizontal fragmentation due to the dominance of small and medium-sized firms facing strong competition, and longitudinal fragmentation resulting from the regular turnover of clients and suppliers across different projects (Hall et al., 2020). In such a fragmented landscape, even advanced technologies tailored for the construction industry may encounter challenges, as both technology developers and users must deal with numerous barriers, including regulatory, commercial, cultural, organisational, and psychological issues. These barriers are central to ethics practices in business organisations. Ethics in the digitalisation of the construction industry is a multifaceted issue influenced by various factors. More so, ethical considerations become crucial as the industry increasingly integrates digital technologies. Ethical issues can impact the integrity and fairness of construction practices regarding safety, privacy, and environmental sustainability. In this section, we will discuss ethical concerns within the context of the construction industry's digitalisation, shedding light on the key challenges organisations face and how to address them.

Poor Data Privacy and Security The use of digital information tools in the construction industry fosters a healthy environment for establishing and expanding organisations that specialise in applying technologies for design and construction (Sacks et al., 2020). However, because of its unique characteristics and reliance on technology, the construction industry has elevated data security to a prominent position. Embracing digital transformation offers substantial advantages and introduces a distinct set of data security challenges. These issues encompass safeguarding sensitive project data, establishing secure communication channels among stakeholders, and regulating access to confidential information. Due to a lack of data security protection, many harmful information breaches occur, drawing increasing attention to building information security concerns. The construction information includes a high number of stakeholders such as clients, contractors, subcontractors, architects, quantity surveyors, and engineers, and the number of construction project files, such as project plans, financial records, intellectual property, and personally identifiable information, is enormous, resulting in a massive volume of information. However, standard network security information protection software is primarily passive, making it difficult to increase its autonomy (Zhao, 2024). Many technologies used in construction capture and handle enormous volumes of data, including personal information on employees, subcontractors, and clients. From remote monitoring to wearable devices, technology continues to improve safety in construction. Monitoring employees is not a recent development; however, technological advancements have made these methods more widespread and possibly more invasive. Ensuring that monitoring technologies enhance safety and productivity while respecting workers' privacy and independence is crucial. Construction companies must follow relevant data protection rules, such as the General Data Protection Regulation (GDPR), which governs the acquisition, storage, use, and sharing of personal information.

Handling vast amounts of data securely and effectively is a significant concern, especially with increasing cyber threats. The risk of system attacks in the construction sector is escalating due to the rising adoption of technology. For example, BIM tools frequently utilise digital resources in construction, yet the issue of data security for BIM tools has not been adequately addressed (Maskuriy, Selamat, Ali, et al., 2019). Data security in blockchains remains vulnerable to data alterations when used in construction projects (Tezel et al., 2020). Research on smart cities in China has revealed the challenges posed by data and system security issues, often complicated by leaks, in various digital technologies (Chen, 2019). The threats imposed by the data and security breach greatly damage client and user's confidence and trust in the entire digital construction process (Koseoglu et al., 2019). As the construction sector continues to evolve digitally, data security is critical, and the industry must prioritise preserving sensitive information, maintaining client trust, and safeguarding its reputation. Overcoming the sector's specific issues requires a complex strategy combining technology, protocols, and people (Zhao, 2024). Ethical handling of data requires secure storage, sound processing, and the establishment of protocols for data access, encryption, and secure sharing to guarantee data integrity and confidentiality and prevent its misuse or exposure.

Transparency Issues For the construction sector to fully harness the chance to reinvent itself in the digital age and move away from past misconceptions and controversies, digital transparency is essential. Transparency allows organisations to leverage data for quicker, more informed decision-making, inspires employees to question existing practices, fosters a deeper comprehension of customer needs, facilitates decentralised decision-making and collaborative creation, and encourages ongoing experimentation, learning, ethical choices, and proactive governance. However, collecting large amounts of data with digital technologies (e.g. sensors, drones, IoT devices) also raises questions about data ownership, consent, and the purpose of usage. Despite the growing trend of digitisation, the construction industry still exhibits a significant lack of transparency and fragmented data and information, which results in poor management of project timelines, costs, and quality throughout the life cycle (Raco et al., 2021). For instance, the widespread use of IoT-sensors by construction companies and the lack of visibility into the control and oversight provided by IoT-based monitoring systems complicate the ability of employees to challenge management decisions derived from sensor-collected data. Some of these digital technologies can obscure the processes and decisions within construction projects, making it harder for stakeholders to understand and scrutinise. To fully digitise the construction process, it is essential to implement information governance that enhances transparency within the organisation. During data integration, data exchange, and supply chain management, maintaining transparency can improve product marketing and ensure long-term sustainability for stakeholders. Processes must be clear, the functions and intentions of digital tools should be conveyed clearly, and decisions—where feasible—should be clarified for those impacted both directly and indirectly.

Accountability and Responsibility Issues Using digital tools in project management, planning, and decision-making can introduce biases that may affect project outcomes and fairness. Decisions made by automated systems can have significant impacts, and it may be unclear who is responsible for those decisions. For instance, through the incorporation of AI applications and tools, certain machines are taking on new roles and, in some cases, gaining a degree of autonomy, along with the expectation to exhibit different types of human intelligence and make judgements regarding workers. However, technologies that automate decision-making can result in unintended or incorrect choices, raising concerns about who holds legal accountability for any harm or negative consequences that may arise (Eurofound, 2023). Also, digitalisation can affect local communities, from displacement due to construction projects to the digital divide impacting local labour forces. The concepts of accountability and responsibility are intricately connected to the ethical principle of fairness. Systems should be established to guarantee that responsibility and accountability regarding digital applications and their results can be assigned before and following their development, deployment, and utilisation (Aghimien et al., 2021). Establishing clear lines of accountability and ensuring human oversight in critical decision-making processes is crucial to upholding ethical standards. Evaluating and addressing adverse effects on local communities while ensuring digitalisation benefits all parties involved, including local workers and residents, is crucial for social responsibility. Keeping clear records and offering stakeholders easy access to information regarding project processes and decisions made through digital tools is essential to ensure accountability.

Unemployment Issues Another ethical concern about digitalisation is the potential for job displacement or changes in job roles, which may affect the livelihood of workers in the construction industry. The digitalisation process speeds up the changes in the labour market and increases the emphasis on flexibility through customised temporary work and service contracts, freelancing work, and holding several jobs (Eichhorst et al., 2017). The change appears to be linked to a subsequent growth in inequality since jobs that require advanced skills and offer high wages are becoming more prevalent (Vasilescu et al., 2020). Hence, with digitalisation revolutionising multiple facets of construction, from design and planning to project management and maintenance, there is a fear of losing jobs in the construction industry. While past research indicated that the likelihood of job losses increases with the digital transformation of organisations (Mzekandaba & Pazvakav, 2018; Watson, 2011), other studies revealed that a shift towards new technologies in work processes may not specifically lead to job replacement but will create a demand for different skill sets (Lawrence et al., 2017; Muro et al., 2017). Integrating digital technologies in construction has created complex digital capabilities that are not easily understood and, more concerning, are mostly utilised by younger individuals, such as construction informatics (Braun & Sydow, 2019). Woodhead et al. (2018) revealed that older and experienced construction workers are displaced with experiential domain knowledge that should be combined with the digital capabilities of young workers. All these are ethical issues that must be addressed.

One aspect of the issue is the potential for digital technologies, like robots, to replace conventional jobs on construction sites that elderly workers often perform, such as material inventories. It is typical in the construction industry to characterise ageing as a lack of contemporary skill (Pan et al., 2020). This approach has hampered digital capacities among older adults and impeded efforts towards digital transformation in the industry (Maskuriy, Selamat, Maresova, et al., 2019). This is not only considered counterproductive but also unethical practice. The right approach should be to assess individuals without bias or set up an institutionalised age management strategy to discover digital skills among older people. Since digital skills are fundamental and the smart labour market will prefer people with digital skills, it is expedient for education and training to be provided according to the needs of the digitised labour market. Developing approaches for retraining and upskilling employees to adjust to emerging technologies guarantees equitable treatment and job opportunities in the changing industry. It is also vital that the financial benefits and productivity improvements from digital transformation leads to equitable pay and better working conditions for employees to uphold ethical labour standards.

Intellectual Property Issues In the modern age of digital change, the importance of intellectual property has grown significantly (Xue et al., 2022). The construction sector generates numerous original works, including architectural designs, software applications, and marketing content. Securing these works with copyright is crucial for preserving exclusivity and stopping unauthorised use, thereby protecting companies' intellectual and creative investments. As digital tools and platforms become integral to construction, issues around intellectual property rights and proprietary technologies arise (Aghimien et al., 2021). Digitalisation often requires collaboration across various stakeholders, raising issues around data sharing and proprietary information. Organisations may use digital technologies protected by intellectual property rights, such as patents, copyrights, and trademarks. While promoting open yet secure collaboration and data-sharing practices, construction companies must respect these rights and obtain appropriate licences before using protected technologies.

Regulatory Compliance and Legal Issues The extensive advancement and implementation of digital tools have turned ethics guidelines and various governance frameworks into a crucial policy focus for numerous regions worldwide. Alongside the USA, China, and Japan, the European Union has prioritised formulating new policies and regulations concerning AI. The development of such regulations varies significantly based on the technology involved. The ethical discussions surrounding blockchain and cloud services appear to be primarily in academic inquiry, whereas the ethics related to robotics and AI are more influenced by governmental guidelines or company self-regulatory measures (Schoentgen & Wilkinson, 2021). As construction companies progressively embrace new technology, it is crucial to comprehend the legal ramifications and guarantee appropriate use to prevent expensive legal conflicts and other adverse outcomes. Digital tools and processes must comply with existing regulations and standards, which can be complex and evolving. Technologies such as AI and IoT can be used in ways that may not yet be fully regulated or understood in the construction industry. Using digital tools in construction can create new risks and uncertainties regarding liability in case of failures or accidents. There are worries that ethical principles are not enforceable, whereas legal standards are. Therefore, a policy strategy that relies only on ethical guidelines without a legal framework might amount to nothing more than a ‘gesture of goodwill’ (Schultz et al., 2024). Ensuring that digital practices align with legal and regulatory frameworks and staying updated on changes is crucial for maintaining ethical and compliant operations. Adopting ethical guidelines and best practices for using emerging technologies, even beyond current regulations, demonstrates a commitment to ethical standards. Implementing robust risk management strategies and clarifying liability in digital processes ensures responsible use and accountability. Construction technology integration can also substantially influence contractual duties and liabilities. Therefore, construction stakeholders should carefully examine legal issues when incorporating technology into their projects and contracts. Construction contracts should explicitly outline the scope of work, including any technology services to be delivered. Construction businesses should also evaluate how liability would be shared during technical breakdowns and how to remain current on industry standards for technology utilisation, such as BIM standards. These standards not only direct the efficient and effective use of technology but also assure legal compliance.

Ethics-Washing The phrase ‘ethics washing’ has been introduced to refer to the practice of manipulating ethics through deceptive communication, giving the illusion of ethical digitalisation without a solid ethical framework, rationale, or the involvement of ethicists (Schultz et al., 2024). It is an approach that self-regulates, incorporating guidelines and codes of digital ethics to provide assurance to the public and lawmakers regarding their ethical behaviour (Hao, 2019). Nevertheless, significant concerns persist regarding the inconsistency between corporate communication on digital ethics, such as ethical AI, and actual business practices (Benkler, 2019). It often stems from an unwillingness to address ethical matters or concerns related to public relations, where the primary goal is to manage the company’s reputation risk. Such methods can seem disingenuous, especially when employed by multimillion-dollar tech firms. Their efforts to seem genuinely concerned about ethical concerns might serve as a strategy to resist potential new regulations by demonstrating to customers and regulators that their products or services are trustworthy. There is a common belief that handling ethical issues might clash with innovation and competitiveness. However, it appears that tech consumers and users are increasingly aware of these matters, and companies would benefit from acting ethically in this regard to stay competitive. Ethical concerns arising from digitalisation are summarised in Table 5.1.

Table 5.1 Summary of ethical concerns in related digital technologies

Ethical concern	Examples of technologies affected
Data ownership	IoT, BIM, Big data
Data control	BIM, IoT, Cloud services
Privacy and data governance	IoT, BIM, AI, Big Data Analytics (BDA)
Security and data protection	BDA, Cloud services, IoT, BIM
Trust	BIM, AI, Robotics
Legal status	Robots, drones, IoT, AI
Decision ability	Robots, AI
Accountability	IoT, BIM, BDA
Societal and environmental well-being	IoT, Robots
Diversity	Robots, AI, BDA
Fairness	Robots, AI, BDA

Source: Author's compilation (2024)

Impact of Ethics on Digitalisation

Given the complex and varied challenges facing the industry, construction companies must make decisions about technology adoption based on a significant amount of trustworthy and high-quality information that promotes ethical practices (Hatoum & Nassereddine, 2024). Utilising such information can help decrease uncertainty, lessen the chances and severity of risks, and boost decision-makers' confidence by improving their understanding of possible outcomes (Dziadosz & Kończak, 2016). This is especially vital for decisions regarding technology adoption, as embracing new technologies involves inherent risks and difficulties that could compromise not only the success of the implementation but also the potential benefits of the technologies being utilised (Taherdoost, 2021). Consequently, it is crucial for decision-makers to evaluate how well the technologies being considered align with the organisation's business from ethical perspectives prior to proceeding with adoption.

There has been a notable increase in the number of ethical frameworks concerning ICTs in recent years, especially in relation to AI and data privacy (Schoentgen & Wilkinson, 2021). Advocates of an ethical perspective argue that ethics offers a structure for evaluating the purpose, design, and societal impact of technology. Conversely, some detractors believe that these ethical frameworks serve as a 'soft' alternative to formal regulations, supporting self-regulatory measures (Wagner, 2018). A report for the European Parliament emphasises that ethics cannot simply be equated with codes of conduct, guidelines, or principles alone; instead, it should be viewed as an ongoing process. The European Parliamentary Research Service contend that the ethical approach involves identifying ethical dilemmas and determining the necessary governance mechanisms, whether that be 'soft' frameworks or 'hard' regulations—to address emerging digital ethical concerns timely.

Improving Ethical Practices in Digitalisation

The pace of technological advancement is projected to increase, given the ability of digital tools to improve various facets of construction tasks. Nevertheless, a significant challenge lies in utilising these tools ethically and responsibly. For future applications of digital tools to genuinely generate value within the construction industry and beneficially influence society, ethical considerations should be the foundation. An essential approach is to formulate an ethical and responsible framework to direct how construction experts can generate ethical value by utilising digital tools. This approach is not limited to a social science examination of construction methods and their decision-making frameworks, internal organisational hierarchies, the regulatory environments they function in, and broader societal elements and ethical aspects that influence the use of digital tools; it is also not just a technical endeavour focused on designing algorithms or programming the code that enables these tools to work, nor is it merely an assessment of the technological attributes of specific tools before they are chosen and utilised. Undoubtedly, creating a responsible and ethical strategy for digitalising the construction sector involves both social and technical aspects that are deeply interconnected. Moreover, employing various digital tools is central to a thorough exploration of real-world practices and in-depth interdisciplinary research that spans the fields of information technology, engineering, construction management, decision-making, as well as ethics and philosophy.

Summary

Digitalisation in the construction industry offers significant benefits but also brings a host of ethical challenges. Addressing these factors requires a proactive and comprehensive approach, integrating ethical considerations into all aspects of digital practice. By fostering transparency, protecting data, promoting fair labour practices, and ensuring environmental and social responsibility, the construction industry can navigate the complexities of digital transformation ethically and sustainably. The chapter concentrated on crucial issues such as data privacy, security, transparency, and the influence on workers. It was revealed that integrating new technology involves capturing and managing massive volumes of data, generating questions about how this information is stored, shared, and used. Ethical concerns about data ownership and potential misuse or breaches necessitate strong frameworks and policies to protect stakeholder interests. The chapter, through bibliometric analysis, revealed that digital technologies, including AI, machine learning, and IoT, had been evaluated for their ethical usage and influence on decision-making. The ethical use of emerging digital technologies on the health and well-being of people, as well as ethical human-computer interactions, have also been explored. While digitisation provides unprecedented prospects for efficiency and innovation in construction, its ethical consequences must be properly controlled in such a way that stakeholders will commit to ethical policies that protect data integrity, promote openness, and support the workforce. Therefore, by reviewing current practices and proposing guidelines for ethical digitalisation, the chapter contributes to the sustainable and responsible evolution of the construction industry. It emphasises the necessity of establishing a balanced strategy in which technological advancement is matched with ethical principles, ensuring that progress benefits all stakeholders while maintaining public trust. The next chapter explored existing theories, frameworks, and models for sustainable ethical practice in the digital era.

References

Aghimien, D., Aigbavboa, C., Meno, T., & Ikuabe, M. (2021). Unravelling the risks of construction digitalisation in developing countries. *Construction Innovation, 21*(3), 456–475.

Aghimien, D. O., Aigbavboa, C. O., Oke, A. E., & Thwala, W. D. (2020). Mapping out research focus for robotics and automation research in construction-related studies: A bibliometric approach. *Journal of Engineering, Design and Technology, 18*(5), 1063–1079.

Akinci, B., & Anumba, C. (2008). Sensors in construction and infrastructure management. *Journal of Information Technology in Construction, 13*(5), 69–70.

Ashok, M., Madan, R., Joha, A., & Sivarajah, U. (2022). Ethical framework for artificial intelligence and digital technologies. *International Journal of Information Management, 62*, 1.

Benkler, Y. (2019). Don't let industry write the rules for AI. *Nature, 569*(7754), 161–162.

Boatright, J. R. (2007). Reluctant guardians: The moral responsibility of gatekeepers. *Business Ethics Quarterly, 17*, 613–632.

Braun, T., & Sydow, J. (2019). Selecting organisational partners for interorganizational projects: The dual but limited role of digital capabilities in the construction industry. *Project Management Journal, 50*(4), 398–408.

Bryson, J., & Winfield, A. (2017). Standardising ethical design for artificial intelligence and autonomous systems. *Computer, 50*(5), 116–119.

Burr, C., Taddeo, M., & Floridi, L. (2020). The ethics of digital Well-being: A thematic review. *Science and Engineering Ethics, 26*(4), 2313–2343.

Bynum, T. W. (2006). Flourishing ethics. *Ethics and Information Technology, 8*, 157–173.

Chen, X. (2019). The development trend and practical innovation of smart cities under the integration of new technologies. *Frontiers of Engineering Management, 6*(4), 485–502.

Chiodo, S., Kaiser, D., Shah, J., & Volonté, P. (2024). What improving technology through ethics means. In *Improving technology through ethics* (pp. 1–10). Springer.

Crane, A., & Matten, D. (2010). *Business ethics: Managing citizenship and sustainability in the age of globalisation*. Oxford University Press.

De George, R. T. (2006). The relevance of philosophy to business ethics: A response to Rorty's "is philosophy relevant to applied ethics?". *Business Ethics Quarterly, 16*, 381–389.

DeLapp, K. M. (2011). *Meta-ethics. Internet encyclopedia of philosophy*. Retrieved September 3, 2024, from https://iep.utm.edu/

Dwivedi, Y. K., et al. (2023). So what if ChatGPT wrote it? Multidisciplinary perspectives on opportunities, challenges, and implications of generative conversational AI for research, practice and policy. *International Journal of Information Management, 71*, 1–11.

Dziadosz, A., & Kończak, A. (2016). Review of selected methods of supporting decision-making process in the construction industry. *Archives of Civil Engineering, 62*(1), 111–126.

Eichhorst, W., Hinte, H., Rinne, U., & Tobsch, V. (2017). How big is the gig? Assessing the preliminary evidence on the effects of digitalisation on the labor market. *Management Revue, 28*, 298–318.

Eurofound. (2023). *Ethical digitalisation at work: From theory to practice*. Publications Office of the European Union.

Fisher, C., & Lovell, A. (2009). *Business ethics and values: Individual, corporate and international perspectives*. Pearson Education.

Floridi, L., & Taddeo, M. (2016). What is data ethics? *Philosophical Transactions of the Royal Society: Mathematical Physical and Engineering Sciences, 374*, 20160360. https://doi.org/10.1098/rsta.2016.0360

Hall, D. M., Whyte, J. K., & Lessing, J. (2020). Mirror-breaking strategies to enable digital manufacturing in silicon valley construction firms: A comparative case study. *Construction Management and Economics, 38*(4), 322–339.

Hallinger, P., & Kovačević, J. (2023). Applying bibliometric review methods in education: Rationale, definitions, analytical techniques, and illustrations. In *International encyclopedia of education* (pp. 546–556). Elsevier.

Hao, K. (2019). IQ 2020, OeWşV VWRS AI eWhLcV-washing and actually do something. *MIT Technology Review*. Retrieved August 12, 2024, from https://www.technologyreview.com/s/614992/ai-ethics-washing-time-to-act/

Hatoum, M. B., & Nassereddine, H. (2024). Determining decision-making factors for technology adoption in the construction industry. In *Industry 4.0 transformation towards industry 5.0 paradigm - challenges, opportunities and practices*. IntechOpen. https://doi.org/10.5772/intechopen.1004365

Hosseini, M. R., Martek, I., Banihashemi, S., Chan, A. P., Darko, A., & Tahmasebi, M. (2020). Distinguishing characteristics of corruption risks in Iranian construction projects: A weighted correlation network analysis. *Science and Engineering Ethics, 26*, 205–231.

Hursthouse, R. (2013). Normative virtue ethics. *Ethica, 2013*, 645.

Khodabakhshian, A. (2024). Ethics-aware application of digital technologies in the construction industry. In *Improving technology through ethics* (pp. 49–64). Springer.

Koseoglu, O., Keskin, B., & Ozorhon, B. (2019). Challenges and enablers in BIM-enabled digital transformation in mega projects: The Istanbul new airport project case study. *Buildings, 9*(5), 115.

Lawrence, M., Roberts, C., & King, L. (2017). *Managing automation: Employment, inequality and ethics in the digital age*. Institute of Public Policy Research. Retrieved August 12, 2024, from www.ippr.org/publications/managing-automation

Li, Y., Sun, H., Li, D., Song, J., & Ding, R. (2022). Effects of digital technology adoption on sustainability performance in construction projects: The mediating role of stakeholder collaboration. *Journal of Management in Engineering, 38*(3), 04022016.

Lu, W., Huang, G. Q., & Li, H. (2011). Scenarios for applying RFID technology in construction project management. *Automation in Construction, 20*(2), 101–106.

Maskuriy, R., Selamat, A., Ali, K. N., Maresova, P., & Krejcar, O. (2019). Industry 4.0 for the construction industry—How ready is the industry? *Applied Sciences, 9*(14), 2819.

Maskuriy, R., Selamat, A., Maresova, P., Krejcar, O., & David, O. O. (2019). Industry 4.0 for the construction industry: Review of management perspective. *Economies, 7*(3), 68.

McCarthy, S. F. (2012). Developing an Australian code of construction ethics. *Australasian Journal of Construction Economics and Building, 12*(2), 100–110.

McDonald, G. M. (2004). A case example: Integrating ethics into the academic business curriculum. *Journal of Business Ethics, 54*(4), 371–384.

Muñoz-La Rivera, F., Mora-Serrano, J., Valero, I., & Oñate, E. (2021). Methodological-technological framework for construction 4.0. *Archives of Computational Methods in Engineering, 28*, 689–711.

Muro, M., Liu, S., Whiton, J., & Kulkarni, S. (2017). *Digitalisation and the American workforce*. Brookings Institute.

Mzekandaba, S., & Pazvakav, R. (2018). *Tech to contribute to job losses in SA*. Retrieved March 11, 2024, from www.itweb.co.za/content/o1Jr5qxEE5YvKdWL

Oladinrin, O. T. (2016). *Implementation of codes of ethics within construction organisations in Hong Kong: Towards an improved ethical behaviour*. Hong Kong Polytechnic University.

Oladinrin, O. T., & Ho, C. M. F. (2016). Embeddedness of codes of ethics in construction organisations. *Engineering, Construction and Architectural Management, 23*(1), 75–91.

Pan, M., Linner, T., Pan, W., Cheng, H., & Bock, T. (2020). Structuring the context for construction robot development through integrated scenario approach. *Automation in Construction, 114*, 103174.

Patil, G. (2019). Applications of artificial intelligence in construction management. *International Journal of Research in Engineering, 32*(3), 32–1541.

Petersen, T. S., & Ryberg, J. (2019). Applied ethics. In *Oxford bibliographies*. Oxford University Press.

Poon, J. (2004). The study of ethical perceptions of construction managers. In *Proceedings of the 20th annual Association of Researchers in construction management (ARCOM) conference*.

Powell, A. B., Ustek-Spilda, F., Lehuedé, S., & Shklovski, I. (2022). Addressing ethical gaps in "technology for good": Foregrounding care and capabilities. *Big Data and Society, 9*, 1–12.

Raco, F., Stefani, M., Balzani, M., & Ferrari, L. (2021). Towards effective project documentation, transparency, and data-driven decision-making through BIM-blockchain based applications. *The International Archives of the Photogrammetry, Remote Sensing and Spatial Information Sciences, 46*, 437–444.

Saarikko, T., Westergren, U. H., & Blomquist, T. (2020). Digital transformation: Five recommendations for the digitally conscious firm. *Business Horizons, 63*(6), 825–839.

Sacks, R., Eastman, C., Lee, G., & Teicholz, P. (2018). *BIM handbook: A guide to building information modelling for owners, designers, engineers, contractors, and facility managers*. John Wiley & Sons.

Sacks, R., Girolami, M., & Brilakis, I. (2020). Building information modelling, artificial intelligence and construction tech. *Developments in the Built Environment, 4*, 100011.

Schoentgen, A., & Wilkinson, L. (2021). Ethical issues in digital technologies. In *Digital societies and industrial transformations: Policies, markets, and technologies in a post-Covid world.* International Telecommunications Society (ITS).

Schultz, M. D., Conti, L. G., & Seele, P. (2024). Digital ethicswashing: A systematic review and a process-perception-outcome framework. *AI and Ethics, 5*, 805–818.

Shaw, W. (2017). *Business ethics* (9th ed.). Cengage Learning.

Stahl, B. C. (2021). *Artificial intelligence for a better future: An ecosystem perspective on the ethics of AI and emerging digital technologies*. Springer.

Suen, H., Cheung, S.-O., & Mondejar, R. (2007). Managing ethical behaviour in construction organisations in Asia: How do the teachings of confucianism, taoism and buddhism and globalisation influence ethics management? *International Journal of Project Management, 25*(3), 257–265.

Taherdoost, H. (2021). Importance of technology acceptance assessment for successful implementation and development of new technologies. *Global Journal of Engineering Sciences*. https://doi.org/10.33552/GJES.2019.01.000511

Tezel, A., Papadonikolaki, E., Yitmen, I., & Hilletofth, P. (2020). Preparing construction supply chains for blockchain technology: An investigation of its potential and future directions. *Frontiers of Engineering Management, 7*, 547–563.

Thompson, D. F. (2005). *Restoring responsibility: Ethics in government, business, and healthcare* (Vol. 575). Cambridge University Press.

Turk, Ž., & Klinc, R. (2017). Potentials of blockchain technology for construction management. *Procedia Engineering, 196*, 638–645.

Vasilescu, M. D., Serban, A. C., Dimian, G. C., Aceleanu, M. I., & Picatoste, X. (2020). Digital divide, skills and perceptions on digitalisation in the European Union—Towards a smart labour market. *PLoS One, 15*(4), e0232032.

Vee, C., & Skitmore, C. (2003). Professional ethics in the construction industry. *Engineering, Construction and Architectural Management, 10*(2), 117–127.

Wach, K., Duong, C. D., Ejdys, J., Kazlauskaitė, R., Korzyński, P., Mazurek, G., Paliszkiewicz, J., & Ziemba, E. (2023). The dark side of generative artificial intelligence: A critical analysis of controversies and risks of ChatGPT. *Entrepreneurial Business and Economics Review, 11*(2), 7–30.

Wagner, B. (2018). Ethics as an escape from regulation: From "ethics-washing" to ethics-shopping? In E. Bayamlioğlu, I. Baraliuc, & L. Janssens (Eds.), *Being profiled: Cogitas ergo sum. 10 years of profiling the European citizen* (pp. 84–88). Amsterdam University Press.

Waller, B. N. (2011). *Against moral responsibility*. MIT Press.

Watson, A. (2011). Digital buildings—Challenges and opportunities. *Advanced Engineering Informatics, 25*(4), 573–581.

Wempe, J. F. D. B., & Kaptein, M. (2002). *The balanced company: A theory of corporate integrity*. Oxford University Press.

Woodhead, R., Stephenson, P., & Morrey, D. (2018). Digital construction: From point solutions to IoT ecosystem. *Automation in Construction, 93*, 35–46.

Xue, X., Tan, X., Huang, Q., Zhu, H., & Chen, J. (2022). Exploring the innovation path of the digital construction industry using mixed methods. *Buildings, 12*(11), 1840.

Zhao, L. (2024). Construction information security management system based on data sharing algorithm. *Intelligent Decision Technologies, 18*, 2775–2788.

Chapter 6
Ethical Practices Theories, Frameworks, and Models for Sustainable Digitalisation

Introduction

In the quest to develop an ethical practice framework for sustainable construction digitalisation, reviewing existing frameworks and models became critical. This is because a preliminary review showed that an ethical practice framework specifically designed for construction organisations was scarce. So, to develop one, it was imperative to review existing frameworks across diverse fields of study. To this end, the chapter focuses on 11 existing ethics frameworks designed for effective and ethical development and use of emerging technologies. These digital ethics and ethics of technology frameworks were designed to implement diverse technologies with general business, telecommunication, health, social services, voluntary organisations, and education. The review of these frameworks specifically identified the different ethical considerations these existing works have deemed crucial for the ethical use of technologies. This gave an insight into the different ethical practices that construction organisations, practitioners, and stakeholders need to consider to sustain their digitalisation. The chapter ends with a summary of the reviewed frameworks and models. It highlights the most prominent ethical considerations to be used in developing the proposed ethical practice framework for sustainable construction digitalisation.

O. Oladinrin et al., *Ethical Practices for Sustainable Construction Digitalisation*, https://doi.org/10.1007/978-3-032-00151-1_6

Review of Related Digitalisation Ethical Practice Frameworks and Model

The significance of ethical behaviours in our daily activities has led to the development of a myriad of frameworks and models designed to ensure morals and a sense of right and wrong are instilled in every individual. While general ethical frameworks such as virtue ethics, deontology, and utilitarianism have been extensively discussed in Chap. 5, it became essential to explore ethical frameworks that revolve around the development and application of digital technologies. The ethics of technology frameworks were developed to guide the ethical development and use of technological innovations. These frameworks are crucial in addressing the complex moral implications arising from integrating technology into society. Subsequent sections highlight some of these frameworks to develop a conceptual, ethical framework for sustained construction digitalisation.

Deloitte's Technology Trust Ethics (TTE) Framework

In 2022, Deloitte developed the technology trust ethics (TTE) framework to help organisations determine the ethical dimensions of their emerging technology products. This framework encompasses seven pillars with 26 sub-attributes, all bounded by governance and controls, reskilling and education, and regulatory compliance and policy (Deloitte, 2022). The framework proposes that the ethical aspects of organisations' emerging technology products must be:

1. Responsible: Through creating and operating technology in a socially responsible manner. This aspect is measured through humane, common, and social good, sustainability-focused, and value-adding.
2. Accountable: By ensuring the creation and enforcement of policies to apportion responsibility for decision-making. This is measured using ownership, resolvable, and answerable criteria.
3. Robust and reliable: By ensuring that the outputs produced by the emerging technology are consistent and accurate and can withstand errors while quickly recovering from unforeseen disruptions. This construct is measured by the technology's predictability, adaptability, consistency, and accuracy.
4. Safe and secure: It entails protection from unforeseen issues that may cause harm to the user. This can be physical, emotional, environmental or digital. This aspect is measured through user friendliness, user protection, and invulnerability of the technology.
5. Private: This aspect prioritises the user's privacy and ensures data is not used or stored longer than the intended period and stated use. The technology should allow users to opt in or out of sharing their information. The framework measures privacy through autonomy, confidentiality, discretional, and consensual.

6. Transparent and explainable: This is done by ensuring users' understanding of how the technology is being used, especially when making decisions. When decisions are made by emerging technology, they must be easy to understand and open to inspection. The aspect of transparency and explainability is measured in the context of justifiable, interpretable, auditable, and visible.
7. Fair and impartial: Ensure that the developed technology allows for inclusivity, equitable application, access, and outcomes. Accessibility, equitability, inclusivity, and unbiasedness are the key attributes of this construct in the framework.

Ethical AI Development Framework

To address the societal, ethical, and legal concerns associated with rapid advancement in AI, Olorunfemi et al. (2024) propose the ethical AI development conceptual framework for IT systems. The conceptual framework integrates principles from ethics, law, and computer science, focusing on transparency, accountability, fairness, privacy, and security in AI systems. The framework suggested that ethical considerations are embedded at every stage of AI development, ensuring responsible deployment and maintenance (Olorunfemi et al., 2024). Building upon diverse multidisciplinary perspectives, the conceptual framework proposed five key constructs essential for the development and integration of AI into IT systems. These are transparency, accountability, fairness, privacy, and security.

1. Transparency: In ensuring that AI algorithms and decision-making processes are understandable to the users, it is essential to prioritise transparency. Through transparency, AI systems function, the data used, and the reason behind their decisions are made clear to the users. Prioritising transparency will foster trust among users while ensuring responsible system use. Felzmann et al. (2020) have earlier noted that it is essential for developers to make their algorithms comprehensible to non-experts by employing interpretable models, using detailed documentation, and using user-friendly interfaces. This transparency allows for verification of accuracy, fairness, and ethical considerations, encouraging scrutiny that can identify biases, errors, or unintended consequences in AI decisions.
2. Accountability: This allows individuals and organisations to be held responsible for the outcomes of AI-driven decisions (de Laat, 2021; Olorunfemi et al., 2024). Accountability helps address positive and negative consequences that allow people harmed by AI systems to have recourse. This construct proposes that developers, organisations, and individuals must take responsibility for the ethical and societal impacts of AI while putting in place clear accountability mechanisms to identify and rectify ethical issues. This aspect of the ethical AI framework allows individuals to seek redress in cases of harm or discrimination caused by AI. This includes grievance procedures, legal oversight, and remedies to address issues arising from AI decisions.

3. Fairness: This construction is germane to the responsible development and use of AI systems as it emphasises reducing and eliminating bias and discrimination for all demographic groups. Fairness demands equitable treatment and outcomes for individuals irrespective of race, gender, or other protected characteristics. As such, developers are expected to actively identify and mitigate biases in algorithms and datasets through fairness-aware techniques, audits, and inclusive design practices (Nassar & Kamal, 2021; Olorunfemi et al., 2024; Varona & Suárez, 2022).
4. Privacy: Privacy has always been a crucial concern in AI development and deployment due to the need to protect users' personal information from unauthorised access and misuse. By prioritising privacy in the design of AI systems, compliance with data protection regulations, safeguarding individuals' rights to control their personal information, and fostering trust in AI systems can be achieved. It is, therefore, essential for developers of AI systems to implement strong privacy protections throughout the AI life cycle, including anonymisation and encryption of data, access controls, and data minimisation to mitigate risks of unauthorised access and breaches. Furthermore, it is important for these systems and their use of data to comply with relevant data protection regulations. This compliance is crucial for respecting individuals' privacy rights and meeting legal obligations regarding handling personal data (Oakley, 2023; Olorunfemi et al., 2024).
5. Security: Considering that security concerns have been one of the major factors hindering the widespread adoption of most emerging digital technologies, employing strict security measures to maintain the integrity, confidentiality, and availability of AI systems and their data is crucial. Implementing strong protocols to defend against malicious exploitation and adversarial attacks is essential while ensuring system reliability and resilience. This construct of the framework, therefore, expects developers to prioritise security through encryption, authentication, access controls, and conducting regular security audits and vulnerability assessments. Emphasis must be placed on the integrity and reliability of the developed systems, making sure that they function as intended and produce accurate results. Measures to detect and address errors, biases, or attacks that could compromise decision-making integrity must be implemented to ensure effective security in the system (Olorunfemi et al., 2024; Uzougbo et al., 2023).

Ethical Framework for AI and Digital Technologies

Using a comprehensive review of existing works on the ethics of technology, Ashok et al. (2022) proposed a conceptual model for the digital ethical assessment of the use of AI in digital technologies. The model proposed four major ontological domains (i.e. physical, cognitive, information, and governance) of the development of AI. These ontological domains were used to assess 14 ethical implications of AI across seven digital technologies archetypes (i.e. platforms, connectivity, actor-based products, sensor-based data collection, analytical insight generation, analytical interaction, and augmented interaction) that can have organisational and social impact.

The model proposes that:

1. Physical domain: This ontological domain focuses on the device layer. It addresses three key ethical implications: dignity and well-being, safety, and sustainability.
2. Cognitive domain: This domain cuts across the service layer of the developed AI system. Six major ethical implications were identified: intelligence, accountability, fairness, promoting prosperity, solidarity, and autonomy.
3. Information domain: This aspect of the model deals with the developed AI system's content layer, highlighting privacy and security concerns.
4. Governance domain: The governance domain deals with the governance layer of the developed AI system. This aspect considers three ethical implications: regulatory impact, financial and economic impact, and individual and societal impact.

The developed conceptual model also noted that the organisational impact could serve as a mediator for the effect of these four ontological domains on the overall societal impact. Moreover, the characteristics of digital technology archetypes can also moderate the effect of these ontological domains on societal impact (Ashok et al., 2022). Overall, the developed conceptual model significantly contributes to the ongoing debate on the ethical use of AI in digital technologies while providing a wide-ranging model for understanding and addressing ethical challenges. The 14 proposed ethical implications serve as a solid foundation for developing a conceptual framework for ethical construction digitalisation.

Framework for Ethical AI Development in Social Services

The continuous advancement in AI and pervasive usage across diverse sectors has seen the need for ethical development and use become more important than ever before. Seniutis et al. (2024) explored the use of AI in social welfare and service delivery. The study strived to integrate AI into the social service sector ethically. To this end, a framework for ethical AI development was conceptualised for the social service sector in Lithuania. The proposed framework viewed AI applications (natural language processing, computer vision, robotics, and machine learning) in different social service stages, ranging from the design of social services to its evaluation. This application then leads to ethical issues that must be addressed using nine key ethical, practical practices. These are cybersecurity, data quality and service, depersonalisation, explainable AI, human involvement, alternative energy services, content moderation, and monitoring. Employing these ethical practices while integrating AI, organisations will be able to address their security risk, bias in algorithms, privacy issues, opacity, loss of autonomy, societal and environmental well-being, and accountability.

Charity Digital Ethics Framework

Digital ethics, also known as information ethics, has been described as the ethical use of information and technology, as well as the moral principles that shape the interaction between people and the digital world. Froehlich (2020) further described it as the ethical principles and standards that shape how individuals behave and interact in the digital space, including the responsibilities and considerations related to using digital technologies, online platforms, and the internet. With the large amounts of data available to construction organisations in this digital age, digital ethics is concerned not so much with what an organisation cannot do in the digital sphere but with what it 'ought not' to do (Rubens, 2024). As such, the digital ethics frameworks designed for charity organisations propose four major guidelines (i.e. respect trust, take care of digital data, be aware of unintentional unethical behaviour, and cultivate a digital ethics culture) that should shape how organisations use the digital sphere and information.

1. Respect trust: When an organisation gathers and stores client data, it is crucial to consider whether the organisation's data practices betray the trust of its clients. Their data should be used in ways they approve of or expect. Moreover, it is essential to ensure transparency, as individuals can only give informed consent if they understand how their data will be used. If organisations lack transparency, they may gain permission for uses that the clients would not have accepted if fully informed, which is unethical (Rubens, 2024).
2. Take care of digital data: Organisations must understand that personal data is private, and respecting that privacy means appropriately safeguarding it. In doing this, they must maintain data integrity by implementing strong governance and audit procedures to prevent data from being altered, deleted, or stolen (Rubens, 2024).
3. Be aware of unintentional unethical behaviour: Accurate data can be misused unintentionally. Organisations must be on the lookout for such instances (Rubens, 2024).
4. Cultivate a digital ethics culture: Organisations need to consider ethical implications in every digital data analysis they conduct. The outcomes of these ethical assessments should inform their decision-making process, and any practices that violate the organisation's code of digital ethics must be adjusted or discarded (Rubens, 2024).

Telkom's Digital Ethics Framework

Similar to the Charity Digital Ethics framework earlier discussed (Rubens, 2024), Telkom, a South African telecommunication company, developed a digital ethics framework designed to guide the company's workers and representatives on decision-making that are aligned to legal and regulatory requirements as well as respect for human rights and dignity (Telkom, 2023). The framework applies to data collection, processing, and handling and information technology services, processes, or applications that use AI or Machine Learning. The framework was designed to align with five major principles of the organisation: transparency, autonomy, beneficence, non-maleficence, and justice.

1. Transparency: This principle is guided by three key elements: traceable, interactive, and comprehensible. It was noted that data used for any decision-making must come from reliable and verifiable sources, while open communication channels must be used to ensure an interactive approach in their service delivery. Moreover, the user's culture and social background must be considered when presenting and communicating digital products and services. This must be communicated respectfully, and in a way that clear understanding is achieved.
2. Autonomy: This principle of the ethics framework is measured by explainability, literacy, and privacy. The data to be gathered and how it will be used in the algorithmic system must be explained to the system users. Also, when decisions are based on such an algorithm, users must be clearly informed of the origin of such decision and the process taken to achieve it. Moreover, the competency of participants involved in the development of the system must be carefully monitored to ensure digital ethics principles are upheld. Finally, protecting the privacy of users who consented to their data being collected and used must be prioritised.
3. Beneficence: This aspect ensures that the digital product being developed is beneficial and has societal relevance. It is measured by sustainability, security, and responsibility. Designing digital products that can serve current and future generations is essential. This can be achieved by ensuring the data's readability is compatible with future systems. In terms of security, data collection, storage, and use must be done safely. Measures must be implemented to bar unauthorised access to and edit these data. Moreover, those involved in developing and managing digital products must be responsible for their various areas of expertise and understand the ethical implications applicable to their areas.
4. Non-maleficence: This aspect of the framework emphasises the risk assessment of digital products and ensures that they do not cause any form of harm to users. Non-maleficence is measured using three critical components: vis, reliability, controllability, and accountability. All parties involved in digital product development must ensure the product is reliable and available. More so, those involved in product development must ensure that they have control and that authorised human interventions can be conducted when needed. Also, there must be a clear structure that shows those accountable for diverse aspects of product development. Roles and responsibilities should be clearly defined.
5. Justice: This principle is measured by impartiality, equality, and proportionality. Assessing datasets for artefacts that could introduce unfair bias related to culture, personal experiences, religion, or other prejudices is essential. As such, introducing various perspectives and data sources into algorithmic systems is essential. Also, digital products must offer equal access and benefits to all users, as biased data can hinder equitable service delivery. More so, the data gathered must be proportionate to its intended use.

Digital Technology, Artificial Intelligence, and Health Equity Ethical Framework

Prathomwong and Singsuriya (2022) observed that emerging digital technologies and AI have continued to gain popularity in the health sector. Moreover, technological advancement has brought about innovative solutions like eHealth, which allows patients easy access to medical experts while speeding up the rate of medical professionals' training (Moghaddasi et al., 2017). While this is a welcomed advancement, the pervasive use of these technologies can lead to health disparities with ethical implications if not carefully assessed and addressed. To this end, Prathomwong and Singsuriya (2022) proposed an ethical framework for digital technology, AI, and health equity. The framework proposed four major principles, which are 'human dignity'—measured through the non-humanisation, privacy, and autonomy attributes; 'justice'—measured through solidarity, diversity, fairness, and accountability; 'non-maleficence'—measured through privacy and safety; and 'beneficence'—measured through public trust, societal and environmental well-being. Like the Telkom (2023) digital ethics framework, the digital technology, AI, and health equity ethical framework promote the protection of users of emerging digital technologies and the ethical use of these digital tools.

Conceptual Framework for Solving Ethical Issues in Generative AI

From the perspective of Generative AI (Gen AI), which has continued to gain prominence across diverse fields, Zlateva et al. (2024) propose a framework for addressing ethical issues associated with the use of Gen AI. According to Rane (2023), the emergence of Gen AI, such as ChatGPT, has significant implications for the construction industry and is gradually transforming activities across many domains. However, like every technology, Gen AI has its ethical implications, and addressing these ethical issues is necessary for effective usage. To this end, the Zlateva et al. (2024) conceptual framework for solving ethical issues in Gen AI proposed carefully considering 14 activities. These are purpose and contextual clarity; transparency and accountability; data rights and privacy; authenticity protocols; security and robustness; fairness audits; economic and social considerations; public engagement and collaboration; continuous evolution; awareness; evaluation; decision-making; implementation; and review process. Consistent with past frameworks earlier reviewed, this conceptual framework serves as an excellent platform for addressing ethical issues in the development and deployment of Gen AI and by extension, other forms of digital technologies with a view to ensuring responsible implementation that would yield optimum benefits while mitigating potential risks.

Ethical Technology Assessment Checklist

Palm and Hansson (2006) proposed a checklist to assess new technologies' ethical implications. The ethical technology assessment (eTA) checklist was designed to identify the adverse effects of new technologies at the initial stage of development. The checklist was designed as a tool to be used iteratively throughout the development of new technologies to identify negative ethical implications at the early stages. The eTA proposes nine crucial ethical implications that must be critically assessed (Palm & Hansson, 2006). These are:

1. Dissemination and use of information: Emerging digital technologies have revolutionised information sharing and ease of acquiring information. While easier access to information generally has positive effects, it also presents ethical challenges and, as such, must be carefully assessed before the integration of a new technology.
2. Privacy: Privacy is an essential issue with emerging digital technologies. Emerging technologies make gathering individual data easier, and this somehow restricts the privacy of individuals. Camera surveillance has become widespread in urban areas, while technologies like biometric authentication, smart cards, and various monitoring tools allow employers to closely track employees. Evidence suggests that working under constant surveillance can lead to stress, decreased productivity, and employee burnout.
3. Impact on social contact patterns: Communication technologies have transformed how people connect and communicate. There is no denying that other technological advancements, like new building designs and work organisation methods, may also influence social interactions, albeit more indirectly. Evidence shows that increased ease of communication has not necessarily led to stronger connections with the prevalence of digital devices replacing face-to-face communication. It is, therefore, necessary to carefully assess the impact of new technologies on social contact patterns.
4. Sustainability: In integrating new technologies, future generations must be responsible.
5. Control, influence, and power: Technological change has historically altered the distribution of control and influence, particularly in workplaces.
6. Gender, minorities, and justice: New technologies should be assessed in terms of equity in gender as well as their effects on minority groups. Special attention must also be given to individuals with disabilities, who are often significantly impacted by technological changes.
7. International relations: Similarly, the impact of emerging technologies on global relations should be explored. These emerging technologies often reshape relationships between nations; as such, there is a need for greater accessibility to innovations, especially in medical technology and biotechnology. Ethically, it is argued that these advancements should be available to those who cannot afford market prices. Moreover, the issues related to information and power distribution can significantly affect international relations.

8. Human reproduction: In recent times, the potential use of advanced genetic technologies for selective reproduction has sparked widespread public debate, as has the controversial idea of human cloning. Religious authorities often oppose these technologies, arguing that they violate the sanctity of life by allowing humans to manipulate what they see as divine creation. Critics from non-religious perspectives raise concerns about the irreversibility of such interventions and the rights of future generations. These developments challenge our notions of normality, health, and quality of life. Ethical assessments are essential when new technologies have the potential to alter human reproduction.
9. Impact on human values: New technologies are likely to impact values and society. Emerging technologies have revolutionised the way we live, our self-perception and moral values. In biotechnology, sceptics warn that modifying genes could diminish respect for human dignity. So also, the increased accessibility of personal information may undermine our respect for privacy. While assessing the effects of these influences on human values is complex, the topic's urgency makes it a crucial area for research and a primary focus of ethical technology assessments.

Digital Technologies Ethical Concerns Assessment Framework

Schoentgen and Wilkinson (2021) noted that the pervasiveness of emerging digital technologies and their impact on the daily activities of humans calls for a critical assessment of their ethical implications. Fjeld et al. (2020) have earlier noted that ethical concerns in the use of emerging digital technologies revolve around individual and societal impact, and addressing them requires control of technologies by humans and applying ethical values to these digital tools. Based on this understanding, Schoentgen and Wilkinson (2021) proposed a seven-construct digital technologies ethical concerns assessment framework based on their level of ethical concerns. These constructs are:

1. Consent and privacy: This construct cuts across data collection, storage, usage, and sharing between entities and the anonymity of individual data. It was further noted that ethical concerns for consent and privacy are mostly high in applications and platforms, AI, BDA, facial recognition, IoT devices, and robotics.
2. Security: This aspect is centred around protection from online data theft and unauthorised access to and storage of personal information and devices through networks. The ethical concerns in this aspect are high for applications and platforms, IoT devices, robotics, and cloud services.
3. Accuracy: Ensuring accurate data are gathered and processed, as well as the accuracy of the algorithms, is essential. AI, BDA, facial recognition, and distributed ledger technology are all technologies with high ethical concerns in accuracy.

4. Transparency: The purpose and method for the collection of users' data, as well as the means of storing and analysing these data, must be made clear to the users. When the algorithm is used to draw conclusions or make decisions, users must be made aware of the design of the algorithm and its purpose. Key digital technologies with high ethical concerns regarding transparency are applications and platforms, AI, BDA, and facial recognition.
5. Accountability: Organisations/developers must be ready to take responsibility for oversight in data collection, storage, and design. Taking responsibility for mishaps from autonomous robots and AI systems is also essential. Again, applications and platforms, AI, BDA, and facial recognition are mostly affected by this ethical concern.
6. Protection from harm: Technology must be safe and protect users from harm. It must protect vulnerable groups while ensuring safety from inappropriate targeting. Applications and platforms, AI, BDA, and facial recognition are mostly affected by this ethical concern.
7. Inclusion and fairness: Data gathered should be fair towards all demographics, and the data processing and algorithms for decision-making must be inclusive and fair. Again, AI, BDA, and facial recognition are mostly affected by this ethical concern.

Framework for Ethical Learning Technology (FELT)

In the quest for ethical use of emerging learning technologies in diverse sectors, the Association of Learning Technology (ALT) developed the Framework for Ethical Learning Technology (FELT) in 2021. Developed through gathering input and feedback from various stakeholders, FELT was designed to support practitioners using digital technologies for learning, teaching, and assessment as well as institutions and the education technology industry. The framework addresses professionals' different learning technology contexts and should be used alongside ethical policies for research and subject- or industry-specific code of practice (Association of Learning Technology (ALT), 2021). The framework consists of four major constructs as follows:

1. Awareness: This aspect of the framework entails four major areas, which are honouring the autonomy and interests of diverse stakeholders; staying mindful, reflective, and reflexive in your approach; engaging in critical thinking about your practices and the broader environments you can impact; and acknowledging the limits of your knowledge and the potential for unconscious bias (Association of Learning Technology (ALT), 2021).
2. Professionalism: The professionalism dimension has five key areas. These are: exhibit accountable, evidence-based practices; dedicate yourself to continuous professional development and skill enhancement; show integrity and honesty; ensure your practices align with relevant laws and institutional policies; and use

knowledge and research to promote and improve ethical approaches (Association of Learning Technology (ALT), 2021).
3. Values: The aspect of value cut across seven areas, which are: fostering the agency and growth of learners; encouraging fair and equitable treatment to enhance access to learning; creating inclusive and supportive learning environments; embracing diversity as a pathway to innovation; designing services and technologies for broad accessibility; being accountable and ready to explain your decision-making; and maintaining openness and transparency (Association of Learning Technology (ALT), 2021).
4. Care and community: The last construct of the framework centres around care and community with five major areas. These are prioritising self-care and the well-being of others, fostering collegiality and mutual understanding, minimising the risk of harm, acknowledging your responsibilities and influence beyond your institution, and sharing and promoting best practices (Association of Learning Technology (ALT), 2021).

Conversely, FELT is an ideal framework for the ethical use of emerging digital technologies within the construction industry. Construction organisations and professionals can ethically integrate emerging digital technologies into their day-to-day activities through careful awareness, attention to professionalism, clearly defined values and prioritising care and the community.

Summary of Reviewed Ethics of Technology Frameworks

The review of extant framework and assessment models revealed that the ethical implication of technology integration in organisational activities cannot be overlooked. Existing frameworks through interdisciplinary fields set the stage for how organisations should examine the ethical implications of technology development and deployment. They also proffer mitigating measures for possible negative consequences like privacy breaches, discrimination, and a loss of public trust associated with technology deployment. Several frameworks for ethical development and the use of varied digital technologies have been developed across diverse fields like education (Association of Learning Technology (ALT), 2021), health (Prathomwong & Singsuriya, 2022), human resources (Mugumo, 2023), information technology (Farayola & Olorunfemi, 2024; Olorunfemi et al., 2024), telecommunication (Telkom, 2023), multi-function businesses (Deloitte, 2022); there was no evidence of a framework designed to ensure effective and ethical use of digital technologies in construction. This further reinforces the need to explore ethical issues in deploying digital technologies and the digitalisation of construction activities.

Table 6.1 summarises key ethical attributes that must be considered to avoid ethical issues in digital technology deployment. Only attributes that occurred at least twice were included in the table for relevance. Based on the summary table, it is evident that 13 ethical considerations were prominent in existing frameworks. It is

Table 6.1 Summary of key ethical attributes

Sources	Privacy	Security/ safety	Transparency	Accountability/ responsibility	Justice/ Equity/ fairness	Sustainability	Autonomy	Intelligence	Regulatory impact	Individual/ societal impact	Dignity/ well-being	Beneficence	Reliability
Palm and Hansson (2006)	√	√			√	√				√			
ALT (2021)		√	√	√	√		√		√		√		
Schoentgen and Wilkinson (2021)	√	√	√	√	√			√					
Ashok et al. (2022)	√	√		√	√	√	√	√	√	√	√		
Deloitte (2022)	√	√	√	√	√								√
Prathomwong and Singsuriya (2022)	√	√		√	√	√	√			√		√	
Telkom (2023)	√	√	√	√	√	√	√					√	√
Rubens (2024)	√	√	√										
Olorunfemi et al. (2024)	√	√	√	√	√								
Seniutis et al. (2024)	√	√		√		√	√			√			
Zlateva et al. (2024)	√	√	√	√	√	√							
Total	10	11	7	9	9	6	5	2	2	4	2	2	2

Source: Authors' compilation (2024)

imperative to note that some ethical attributes are interlinked and used interchangeably in some frameworks. For instance, some cases used accountability and responsibility as a single construct. Similarly, equity, fairness, and justice have been referred to as related attributes. Looking at the summary in Table 6.1, it can be deduced that the most prominent considerations for effective and ethical integration of digital technologies are security and safety ($f = 11$), privacy ($f = 10$), equity and fairness ($f = 9$), accountability ($f = 9$), transparency ($f = 7$), and sustainability ($f = 6$). Although autonomy emerged four times, in most of these cases (Prathomwong & Singsuriya, 2022; Telkom, 2023), it was noted in relation to privacy. By ensuring users' privacy and making sure they are in control of their decisions and how their information is used, the ethics of autonomy is somewhat achieved. As such, in the context of this study, we adopted privacy and autonomy as a related concept in the ethics of digital technology usage.

Similarly, while it is necessary to carefully consider the individual and societal impact of developing and deploying emerging digital technologies, in the context of this study, emphasis is placed on the societal impact the digitalisation of construction services will have, which is assessed under the sustainability construct. Again, the sustainability performance of many organisations is measured by environmental, social, and governance (ESG) targets (Brounen & Marcato, 2018). This sustainability framework assesses organisational performance based on their impact on the environment through assessment of their contribution to energy efficiency and resource sustainability, societal impact in terms of air labour practices, respecting human rights, and promoting gender diversity, as well as their corporate governance function (Krantz & Jonker, 2024). Considering the importance of ESG, the sustainability dimension of the ethical framework will help understand not just the individual and societal impact of digital technologies but also the environmental impact to ensure ethical integration and digitalisation in the construction industry.

Based on the foregoing, the proposed conceptual, ethical practice framework for sustainable construction digitalisation is rooted in six major ethical considerations. These ethical considerations are

1. Accountability.
2. Equity and fairness.
3. Privacy.
4. Safety and security.
5. Sustainability.
6. Transparency.

Summary

This chapter explores different frameworks designed to aid the development and deployment of digital technologies in diverse sectors. The chapter revealed that despite the importance attribute to ethical concerns of emerging technologies, there is no ethical framework designed to ensure effective and ethical implementation of digital technologies in the construction domain. The chapter concludes that among the arrays of ethical framework and guidelines some ethical considerations were reoccurring which indicates their importance. While some of these ethical considerations were overlapping with others and somewhere not significant, six stood out and were considered for the proposed ethics for sustainable construction digitalisation framework. These are accountability, equity and fairness, privacy, safety and security, sustainability, and transparency. The next chapter of the book presents the conceptualised framework.

References

Ashok, M., Madan, R., Joha, R., & Sivarajah, U. (2022). Ethical framework for artificial intelligence and digital technologies. *International Journal of Information Management, 62*, 1–17.

Association of Learning Technology (ALT). (2021). *Framework for ethical learning technology (FELT).* Retrieved August 12, 2024, from https://www.alt.ac.uk/sites/alt.ac.uk/files/assets_editor_uploads/documents/ALT'%20Framework%20for%20Ethical%20Learning%20Technology%20%28FELT%29%202021.pdf

Brounen, D., & Marcato, G. (2018). *Sustainable insights in public real estate performance: ESG scores and effects in REIT markets*. Berkeley Lab.

De Laat, P. B. (2021). Companies committed to responsible AI: From principles towards implementation and regulation? *Philosophy & Technology, 34*, 1135–1193.

Deloitte. (2022). A framework for ethical tech. Retrieved May 29, 2024, from https://action.deloitte.com/insight/3719/a-framework-for-ethical-tech

Farayola, O. A., & Olorunfemi, O. L. (2024). Ethical decision-making in IT governance: A review of models and frameworks. *International Journal of Science and Research Archive, 11*(2), 130–138.

Felzmann, H., Fosch-Villaronga, E., Lutz, C., & Tamò-Larrieux, A. (2020). Towards transparency by design for artificial intelligence. *Science and Engineering Ethics, 26*(6), 3333–3361.

Fjeld, J., Achten, N., Hilligoss, H., Nagy, A., & Srikumar, M. (2020). *Principled artificial intelligence: Mapping consensus in ethical and rights based approaches to principles for AI.* Berkman Klein Center Research Publication.

Froehlich, T. J. (2020). Ten lessons for the age of disinformation. In K. Dalkir & R. Katz (Eds.), *Navigating fake news, alternative facts, and misinformation in a post-truth world.* IGI Global Publisher. https://doi.org/10.4018/978-1-7998-2543-2.ch003

Krantz, T., & Jonker, A. (2024). *What is environmental, social and governance (ESG)?* Retrieved August 12, 2024, from https://www.ibm.com/topics/environmental-social-and-governance

Moghaddasi, H., Amanzadeh, M., Rahimi, F., & Hamedan, M. (2017). eHealth equity: Current perspectives. *Journal of the International Society for Telemedicine and eHealth, 5*, e9.

Mugumo, V. P. (2023). A systematic approach to ethical technology integration, with a focus on the role of HR professionals and professional bodies. *The People Magazine.* Retrieved August 12, 2024, from https://peoplefactor.co.za/a-systematic-approach-to-ethical-technology-integration-with-a-focus-on-the-role-of-hr-professionals-and-professional-bodies/#:~:text=Ethical%20lapses%20in%20technology%20can%20lead%20to,aligns%20with%20ethical%20values%20and%20societal%20norms

Nassar, A., & Kamal, M. (2021). Ethical dilemmas in AI-powered decision-making: A deep dive into big data-driven ethical considerations. *International Journal of Responsible Artificial Intelligence, 11*(8), 1–11.

Oakley, A. (2023). HIPAA, HIPPA, or HIPPO: What really is the health insurance portability and accountability act? *Biotechnology Law Report, 42*(6), 306–318.
Olorunfemi, O. L., Amoo, O. O., Atadoga, A., Fayayola, O. A., Abrahams, T. O., & Shoetan, P. O. (2024). Towards a conceptual framework for ethical AI development in IT systems. *Computer Science & IT Research Journal, 5*(3), 616–627.
Palm, E., & Hansson, S. O. (2006). The case for ethical technology assessment (eTA). *Technological Forecasting and Social Change, 73*, 543–558.
Prathomwong, P., & Singsuriya, P. (2022). Ethical framework of digital technology, artificial intelligence, and health equity. *Asia Social Issues, 15*(5), 1–14.
Rane, N. (2023). Role of ChatGPT and similar generative artificial intelligence (AI) in construction industry. https://doi.org/10.2139/ssrn.4598258
Rubens, P. (2024). *What is digital ethics?* Retrieved August 12, 2024, from https://charitydigital.org.uk/topics/what-is-digital-ethics-7593
Schoentgen, A., & Wilkinson, L. (2021). Ethical issues in digital technologies. In *Digital societies and industrial transformations: Policies, markets, and technologies in a post-Covid world.* International Telecommunications Society (ITS).
Seniutis, M., Gružauskas, V., Lileikiene, A., & Navickas, V. (2024). Conceptual framework for ethical artificial intelligence development in the social services sector. *Human Technology, 20*(1), 6–24.
Telkom. (2023). *Digital ethics framework.* Retrieved May 29, 2024, from https://group.telkom.co.za/documents/sustainability/Digital_Ethics_Framework.pdf
Uzougbo, N. S., Akagha, O. V., Coker, J. O., Bakare, S. S., & Ijiga, A. C. (2023). *Effective strategies for resolving labour disputes in the corporate sector: Lessons from Nigeria and the United States.*
Varona, D., & Suárez, J. L. (2022). Discrimination, bias, fairness, and trustworthy AI. *Applied Sciences, 12*(12), 5826.
Zlateva, P., Steshina, L., Petukhov, I., & Velev, D. (2024). A conceptual framework for solving ethical issues in generative artificial intelligence. In A. J. Tallón-Ballesteros et al. (Eds.), *Electronics, communications and networks*. IOS Press.

Chapter 7
Conceptualising Effective Ethical Practice for Sustainable Construction Digitalisation

Introduction

The construction industry and its organisations are not immune to the pervasiveness of emerging digital technologies revolutionising how diverse industries operate. To ensure effective integration, there is a need to be conscious of the ethical implications of adopting these digital technologies. Ethics of technology, which guide the development and utilisation of emerging technologies through effective consideration of morality, management of risk and the rights of users, can help in guiding construction organisations through the ethical integration of these technologies (Braunack-Mayer et al., 2012; Dauchess, 2024). Moreover, ethically integrating these technologies can lead to sustainable and continuous use, thereby leading to a more sustainable digitalisation of construction. Therefore, to conceptualise an ethical practices framework for sustainable construction digitalisation, it becomes essential to identify the required digitalisation parameters and ethical considerations needed for construction organisations to consider.

Based on this understanding, the dimensions of construction digitalisation were first explored. Existing digitalisation frameworks across diverse fields were explored, and the key dimensions required by construction organisations to attain digitalisation were unearthed. Next, the six ethical considerations identified from existing literature and presented in the previous chapter were contextualised for the study with their measurement variables identified. The expected outcomes of effectively integrating ethical practices in the digitalisation of construction organisations were also identified, and at the end, a conceptual framework was presented. The proposed ethical practice framework for sustainable construction digitalisation is designed to serve as a platform for further assessment within construction organisations. This implies that when adopting the proposed framework, it is important to test its applicability within the local context and adjust to suitability and applicability as appropriate.

O. Oladinrin et al., *Ethical Practices for Sustainable Construction Digitalisation*, https://doi.org/10.1007/978-3-032-00151-1_7

Constituents of the Conceptualised Ethical Practice Framework for Sustainable Construction Digitalisation

This section presents the constituents of a conceptualised ethics practice framework for sustainable construction digitalisation. The digitalisation dimensions and the ethical considerations that make up the proposed framework are discussed accordingly.

Construction Digitalisation Dimensions

Digitalisation, which is the deployment of digital technologies to improve services and generate new value streams in organisations, has continued revolutionising diverse industries, including construction (Gartner, 2016). With construction organisations designed to offer value to construction clients while making profits, it is unsurprising to see the continuous clamour for digitalisation among construction practitioners, stakeholders, and researchers. Unfortunately, a clearly defined roadmap for the digitalisation of construction organisations is scarce, so there is a need to leverage digitalisation models and frameworks developed in diverse fields.

Aghimien et al. (2021) noted that existing digitalisation models give pointers to the various dimensions wherein construction organisations seeking to be digitalised must place their focus. Gill et al. (2016) developed a framework as a road map for business organisations to achieve digital maturity. The framework explored four critical dimensions (i.e. culture, technology, organisation, and insights). While the aspect of culture relates to a firm's approach towards embracing emerging digital technologies and empowering workers with these technologies, the aspect of technology relates to using these emerging technologies in carrying out specific business functions. The aspect of organisation points to the firm's alignment to digital strategies, governance, and execution, while insight relates to the firm's use of customer and business data to measure success and inform strategy.

Boström and Celik (2017) proposed a digital strategising model to guide business owners. The model proposed six dimensions: communication, value measurement, leadership, ecosystem, technology, and skill. Similarly, Deloitte (2018) proposed a business digitalisation model encompassing five key areas: customer, strategy, technology, operations, and organisation and culture. Aguiar et al. (2019) unearthed similar dimensions. The study proposed that customer, value proposition, organisation, data, operation, and transformation management are key areas that businesses seeking digitalisation should look out for. Likewise, Leyh et al. (2021) proposed six critical areas: corporate organisation, technology, customer, project management, value creation, and proposition. Altimeter (2024) proposed the 'opposite' framework, which considers the organisation's orientation in driving meaningful change, the people, processes, objectives, structure, insights, technology, and execution.

In the manufacturing industry, Gökalp and Martinez (2021) developed a digital transformation capability model that considers strategic governance, which covers organisations' strategy. The model also considered the importance of information and technology, digital process transformation, and workforce management as key capability areas required for organisations to attain digital transformation. In the telecommunication sector, Valdez-de-Leon (2016) presented seven dimensions deemed crucial to the digital transformation of communication service providers. These are technology, strategy, organisation, customer, ecosystem, operations, and innovation. Newman (2017) also proposed technology, strategy, customer, operations, and culture for organisations in the telecommunication sector. In telemedicine, Dyk and Schutte (2012) proposed technology, users, finances, procedures, and policy, while in education, Sheikhshoaei et al. (2018) proposed organisational governance and organisational–human and technical content as key dimensions for digitalisation.

Critical evaluation of these various digitalisation dimensions and their measurement variables show overlaps across existing digitalisation models. While some exist as dimensions in some models, they only exist as measurement variables in others. These overlaps imply a need for commonalities to be identified and applied in the context of the construction industry. Based on this understanding, it is deduced that existing digitalisation models have emphasised the need for technology, people, processes, and strategy. Therefore, exploring these four major capabilities in the context of construction is crucial to the digitalisation of construction organisations.

Technology This dimension refers to all technical attributes that cut across the effective implementation of emerging technologies within construction organisations. This is measured from the perspective of planning and optimising the use of emerging tools such as the Internet of Things (IoT), Big Data Analytics (BDA), Artificial Intelligence (AI), robotics and automation, cloud computing, Building Information Modelling (BIM), blockchain technology, design simulation and visualisation, among others in the delivery of construction services (Bibby & Dehe, 2018; Deloitte, 2018; Leyh et al., 2021; Newman, 2017). It also entails ensuring cybersecurity to protect organisation's key information (Gökalp & Martinez, 2021; Leyh et al., 2021), ensuring effective technology governance through creating principles to guide technology usage (Gökalp & Martinez, 2021; Newman, 2017; Sheikhshoaei et al., 2018), prioritising flexible use of adopted technologies (Boström & Celik, 2017; McLaughlin, 2017) and ensuring adopted technologies aligns with existing organisational resources (Gill et al., 2016). Based on the extant review of digitalisation literature, the following measurement constructs are deemed useful in measuring the technology dimension of construction digitalisation.

1. Adopt analytics and AI (predictive analytics, machine learning, BDA).
2. Adopt automation and robotics (autonomous robots, drones, 3D printing).
3. Adopt collaborative and mobile applications (blockchain).
4. Adopt cybersecurity measures to protect project information.
5. Adopt information management technologies (cloud computing, IoT, data management systems).

6. Adopt sensing and warning-based technologies (smart wearables, safety management software).
7. Adopt simulation technologies (simulation software, Digital Twin).
8. Adopt visualisation technologies (BIM, AR, VR, 3D rendering).
9. Develop principles to guide technology usage.
10. Ensure adopted technology aligns with existing organisational resources.
11. Prioritise flexibility in digital technology deployment.

People The aspect of people encompasses all activities directly related to people, either employees, employers, or customers. Aghimien et al. (2021) noted that every activity within the construction industry is championed by people guided by some specific culture within their organisation. This makes this dimension crucial within the construction context. Based on the review of past models, key variables have been noted. These include technical know-how (Gill et al., 2016), digital knowledge management (Aguiar et al., 2019), reskilling and continuous learning (Deloitte, 2018; Gill et al., 2016; Gökalp & Martinez, 2021; Newman, 2017), support from management (Altimeter, 2024; Gill et al., 2016; Leyh et al., 2021), digital empowerment of workers, digital culture, organisation's attitude to change (Aguiar et al., 2019; Gökalp & Martinez, 2021; Leyh et al., 2021; Newman, 2017), and attracting and retaining digital talents (Sheikhshoaei et al., 2018).

Measuring the people dimension for construction digitalisation entails the following:

1. Create a digital culture within the organisation.
2. Digital management of organisation's knowledge.
3. Digitalisation support from top management.
4. Promote innovativeness among workers.
5. Promote a positive change attitude in the organisation.
6. Put measures in place to attract and retain digital skills.
7. Reskilling of existing organisational workforce.
8. Train workers to develop digital skills.

Process The process dimension cuts across organisational and managerial processes that coordinate activities to deliver better products and services. From a construction perspective, throughout the construction life cycle—from design to execution to operations—digital capabilities must be leveraged. In the design phase, technologies enable data-driven approaches like measurement, interdependency analysis, virtual simulation, and collaborative work. They also improve cost estimation, tendering, and procurement (Sawhney & Knight, 2023). During construction, digital tools facilitate real-time data sharing, integrated planning, automated methods, and rigorous monitoring. They also enable project tracking, deterioration detection, predictive maintenance, and centralised data management—leading to life cycle cost reductions and enhanced structure upkeep (Gerbert et al., 2016). As such, to effectively explore the process dimension of digitalisation, the applicable project management knowledge areas such as digital planning, cost, schedule, quality, scope, resource, risk, change, communication and stakeholder management, and integrated services are worth exploring (Boström & Celik, 2017; Newman, 2017; Project Management Body of Knowledge (PMBOK), 2017; Valdez-de-Leon, 2016).

Based on the foregoing, the process dimension of construction digitalisation can be measure through the following:

1. Digitalise procurement process.
2. Digitalise the project management process (including cost, schedule, resource, quality, risk, change, etc.).
3. Deploy digital technologies in the operation phase of construction projects.
4. Employ applicable digital technologies in the construction process.
5. Prioritise technologies for attaining sustainability in construction.

Strategy This dimension can be traced to the dynamic capability theory (Teece, 2010), which involves an organisation's ability to sense and seize digitalisation opportunities while reconfiguring its resources and activities to promote digitalisation. The strategy dimension takes into consideration key variables such as forward-thinking research and development (R&D) to ensure careful digital technology selection (Teece, 2010), the ability to forecast client needs (Aguiar et al., 2019; Day, 2011), market selection and service boundary (Teece, 2010), evaluation of feedbacks to make informed decisions (De Carolis et al., 2017), investment in digital tools (Gökalp & Martinez, 2021; Newman, 2017) measurement of the performance of adopted digital approaches (Teece, 2010), careful alignment of digital resources with business (Newman, 2017), digital risk culture and open innovation (Teece, 2010). Based on the foregoing, the strategy dimension of construction digitalisation can be measured through:

1. Careful assessment of the business environment and market demand.
2. Careful digital technology selection to align with business objectives.
3. Evaluate clients' feedback using applicable technologies.
4. Investment in carefully selected digital technologies.
5. Investment in forward-thinking R&D.
6. Measurement of the performance of adopted digital technologies for continuous improvement.
7. Promote open communication and effective information management.
8. Use applicable technologies to envisage clients' changing needs.

As seen in Fig. 7.1, these four dimensions form the basis upon which ethics must be considered by construction organisations seeking to be digitalised. These organisations need to prioritise the ethical considerations earlier identified in Chap. 6 (i.e. accountability, equity and fairness, privacy, safety and security, sustainability, and transparency) within these four digitalisation dimensions for effective and ethical deployment of emerging digital technologies for attaining sustainable construction digitalisation.

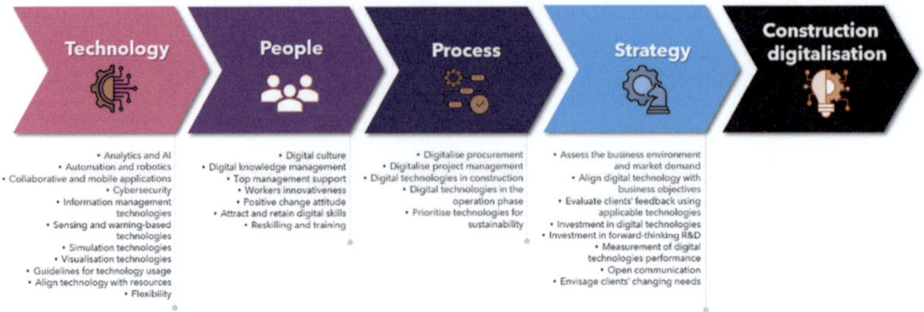

Fig. 7.1 Construction digitalisation dimension. Source: Author's compilation (2024)

Ethical Considerations

The ethical considerations adopted for the conceptualised ethical practice framework for sustainable construction digitalisation are those unearthed from the review of existing technology ethics frameworks in the previous chapter.

Accountability It has been noted that since machines cannot make ethical decisions nor do they have a sense of moral values, humans must remain the moral agents and be accountable for every decision taken through the deployment of these technologies (Institute of Chartered Accountants in England and Wales, 2019). As such, De Laat (2021) noted that humans are responsible for the outcomes of technology-driven decisions. Similarly, Schoentgen and Wilkinson (2021) submitted that when organisations employ emerging technologies, they must take responsibility for any issues that might arise from using such technology. This implies that in the quest for digitalisation, construction organisations must be ready to put policies in place to ensure decision-making responsibilities are apportioned as appropriate, as noted by Deloitte (2022). These organisations must be ready to be accountable for the outcomes of the technologies that they have employed.

Past studies have adopted several measurement attributes to measure accountability as an ethical consideration in technology deployment. Deloitte (2022) measured this ethical consideration using ownership of technology and the outcomes thereof, being able to resolve ethical issues and answer queries. It was suggested that organisations put policies in place to clearly define responsibilities and roles. According to Olorunfemi et al. (2024), through accountability driven by clear policies, consequences of technology adoption can be addressed, and in cases of negative impact, those harmed can seek the necessary recourse. In the context of construction, considering the fact that multiple participants are involved in the delivery of projects (Bikitsha et al., 2014), prioritising accountability in the use of digital technologies can go a long way in settling disputes and addressing claims issues should they occur. By employing emerging technologies such as cloud computing, BDA, and IoT to gather and analyse data on construction projects, construction organisations must put measures in place to ensure that these data are used ethically. They must also take responsibility for any negative individual or societal impact that might arise from collecting and using these data (Olorunfemi et al., 2024; Schoentgen & Wilkinson, 2021).

This ethical consideration is crucial to digitalisation's technology, people, policy, and strategy dimensions. For the technology dimension, prioritising accountability implies that construction organisations put policies in place that will help in the effective planning and selection of appropriate digital technologies that align with their organisational goals and meet their client's objectives. Moreover, people drive construction, and their impact on the successful outcome of construction projects has been noted in existing studies (Aghimien & Aigbavboa, 2024). Accountability will ensure that individuals are assigned roles and use technologies based on their competence and are held liable for the outcomes of using these technologies. Furthermore, when issues occur due to the use of agreed digital technologies on a project, the extent of liability among the different parties involved in the project is clearly outlined in the policy. Also, since construction involves a series of processes, prioritising accountability in using emerging technologies across these processes is essential. For instance, in risk management, accountability in the use of digital technologies can help ensure the right individuals are responsible for using these technologies based on their expertise. This reduces the risk of poor project delivery.

Based on the foregoing, accountability in the context of this study is conceptualised as 'having clearly defined guidelines detailing the responsibilities and liabilities of various parties involved in a construction project in the deployment of emerging technologies'. This ethical consideration is measured through:

1. Having guidelines that define ownership of responsibilities and liabilities.
2. Having mechanisms to identify and resolve ethical issues.
3. Having guidelines for seeking redress in cases of harm or negative impact.

Equity and Fairness Equity and fairness have been attributed to justice in some ethics of technology frameworks (Palm & Hansson, 2006; Prathomwong & Singsuriya, 2022; Telkom, 2023). Palm and Hansson (2006) have earlier noted that new technologies need to be assessed based on their equity in gender as well as their effects on minority groups to ensure no discrimination exists. Deloitte (2022) noted that assessing ethics from the point of fair and impartial application of technologies and the outcomes is essential. This aspect of ethical consideration was measured using accessibility, equitability, inclusivity, and unbiasedness. Other studies have noted that to achieve equity and fairness in new technologies, organisations must put measures in place to identify and mitigate biases in datasets through fairness-aware techniques, audits, and inclusive design practices (Olorunfemi et al., 2024; Varona & Suárez, 2022). Schoentgen and Wilkinson (2021) also emphasise the need for inclusion and fairness by noting that data used to make informed decisions must be fair to all demographics and that the analysis process must be one that allows inclusivity.

For construction organisations seeking digitalisation, prioritising equity and fairness in using digital technologies can significantly influence the ethical use of these technologies. Concerted efforts must be made at the strategy dimension of construction digitalisation to create a digital culture that promotes fairness and equity in using digital tools for services and decision-making in construction organisations. The technology employed must not create a divide between project participants but must allow for inclusivity of all parties involved in the project. More so, when used in selecting project participants or collaborators, fairness and inclusivity in the selection process must be prioritised.

Based on the above knowledge, this study conceptualised equity and fairness in the context of ethical construction digitalisation as 'putting measures in place to ensure that digital technologies employed are inclusive and fair to all participants affected by the delivered construction project and that decisions made based on the results of these technologies do not exhibit bias against any demographic groups'. This ethical consideration is measured through:

1. Carefully select digital technologies that all project participants can understand.
2. Create guidelines to audit fairness in decisions made using digital technologies.
3. Ensure equitable outcomes from technologies used for all project participants.
4. Prioritise inclusivity in the analysis process of data used to make informed decisions.
5. Promote inclusive design practices through the use of digital technologies.

Privacy Privacy has been described as a fundamental human right central to the ethics of technology. It empowers individuals to control their personal information and protect themselves from unwanted intrusion or exploitation (Zostant & Chataut, 2023). Deloitte (2022) emphasised the importance of ensuring user privacy is protected. It was noted that users should always have the option to opt in or out of sharing their information, and whatever information they choose to share is not used or stored longer than the intended period and stated use. To achieve this, Olorunfemi et al. (2024) noted that there must be compliance with existing regulations that protect individuals' rights to control their personal information. Moreover, Rubens (2024) noted that it is essential for organisations to maintain data integrity through strong governance and audit procedures designed to prevent data from being altered, deleted, or stolen. To attain privacy in the ethical use of new technologies, Deloitte (2022) noted that there must be autonomy, confidentiality, discretion, and consent. From the perspective of autonomy in data usage, it is imperative to clearly explain to users the type of data that will be gathered and how it will be analysed and used. In the same vein, it is crucial to assess the competency of participants involved in using these technologies to process the data gathered to ensure they understand the need for privacy (Deloitte, 2022; Telkom, 2023).

Since construction projects generate lots of data, using digital technologies to analyse these data must prioritise privacy where applicable. Using appropriate technologies that ensure data are effectively stored, analysed and destroyed as appropriate is germane to the ethical use of digital technologies. As such, this study conceptualised privacy in the context of ethical construction digitalisation as 'having clear guidelines for the ethical gathering, storing, analysing and destroying of data gathered through the use of digital technologies while giving project participants autonomy of their personal information and how they want such information to be used'. This ethical consideration is measured through:

1. Align technology adoption with existing regulations that protect individuals' rights to control their personal information.
2. Be familiar with existing legal frameworks regarding the handling of personal data.
3. Information harnessed and analysed are kept confidential.
4. Invest in data security to avoid unauthorised alteration, deletion, or theft.
5. Prioritise seeking consent to gather, store, and analyse individuals' data.
6. Promote autonomy of personal information.
7. Have guidelines to clearly inform project participants of the type of data to be gathered and how such data will be used.

Safety and Security Safety and security have continued to emerge as a topical concern in the ethical development and deployment of emerging technologies. This entails putting protective measures in place to address unforeseen issues that may cause harm in the use of technology (Deloitte, 2022). Importantly, measures to ensure the integrity and confidentiality of the data gathered are crucial (Olorunfemi et al., 2024). Schoentgen and Wilkinson (2021) viewed security as protecting online data from theft and unauthorised access to and storage, and they believe that it must be given the utmost priority in technology adoption.

In construction, blockchain's cryptographic features afford optimum security of sensitive operational data and transactions, thereby protecting them from unauthorised access (Sahu, 2023). Also, BIM can help in assessing risks inherent in data sharing and collaboration when the BIM platform is used. This can help reduce the impact of these risks, as early detection can allow adequate mitigation measures to be put in place (Morin & Romero-Torres, 2024). Since one of the major challenges facing the construction industry is the shortage of technical skills to handle emerging digital technologies (Oke et al., 2018), training and creation of awareness on secure communication practices and how to recognise security threats will go a long way in ensuring ethical use of these technologies. When implementing technologies such as drones for capturing data on sites, access controls can restrict operators and the type of data gathered (Hussain et al., 2022). Furthermore, when IoT and sensors are used with construction equipment to gather real-time data, regular firmware updates must be carried out to prevent security flaws and malicious access. Also, these devices must be regularly monitored to detect unusual activities and possible breaches (Struk, 2024).

Based on the foregoing, safety and security in the context of this study is conceptualised as 'the protection of individuals and digital assets from unforeseen activities such as unauthorised access, data breaches, and cyber threats that might cause harm'. This ethical consideration is measured through:

1. Adopt digital technologies to protect data at all stages of the construction project.
2. Careful selection of service providers.

3. Ensure compliance with existing best practice frameworks and regulations.
4. Ensure training of workers on the security of data from adopted technologies.
5. Have guidelines to monitor security measures and remedy issues found.
6. Invest in encryption, authentication, and access controls.
7. Use multi-factor authentication.

Sustainability Adopting new technologies poses environmental, social, and financial impacts (Ashok et al., 2022). Employing sustainability strategies to minimise environmental impact while optimising the efficiency of resources and social responsibility is paramount for the effective and ethical use of emerging digital technologies. Organisations need to employ technology in a socially responsible way by considering humane, common and social good (Deloitte, 2022). Moreover, Telkom (2023) proposed that the benefit of technology should be evident in its sustainability, while Palm and Hansson (2006) proposed that when deploying new technologies, there must be responsibility for future generations.

In construction, BIM can be employed to assess the environmental impact of methods and materials throughout the life cycle of a project. By employing BIM for life cycle assessment, informed decisions can be made, and project participants are aware of the sustainability impacts of their decisions, thereby promoting accountability (Eze et al., 2022; Lemperos, 2024; Marrero et al., 2020). Moreover, IoT and sensors such as light, occupancy, air quality, leakage, temperature, and humidity sensors can be used to ensure energy efficiency in construction projects, demonstrating a commitment to the global agenda on reducing carbon emission in construction (Milesight, 2024). Also, since construction has been known for its high generation of waste (Aghimien et al., 2018; International Energy Agency, 2019; World Green Building Council, 2019), deploying technologies like digital twin and 3D printing can address this issue of waste generation. Digital twins will help simulate the construction process for precise planning to reduce waste, while 3D printing will help reduce waste through the on-demand creation of construction components (Aghimien et al., 2020). By doing this, harm to the environment is eliminated, and natural resources are utilised efficiently. Ethical sourcing and supply chain management can be achieved through blockchain technologies and other software. Blockchain will provide greater traceability in the supply chain and ensure materials meet sustainability criteria.

Based on the foregoing, sustainability in the context of this study is conceptualised as 'having guidelines that consider the environmental, social and economic impact of digital technology deployments while ensuring that these technologies are used responsibly without creating issues for future generations'. This ethical consideration is measured through:

1. Adhere to sustainability frameworks and best practices.
2. Consider the social implications of the selected digital technology.
3. Deploy technology in sourcing and supply chain management.
4. Employ technologies that encourage waste reduction.
5. Employ technologies that promote energy efficiency.

6. Employ technology to assess the environmental impact of materials and methods throughout the project life cycle.
7. Promote awareness and training on the importance of attaining sustainability through adopted technologies.

Transparency It is important to ensure that all participants easily understand decisions made using emerging technologies and must be open to scrutiny. Moreover, the data gathered, the reason behind their gathering, and the storage and analysis system must be clear to the participants (Deloitte, 2022; Olorunfemi et al., 2024). Ensuring that the analysis's outcome is comprehensible, interpretable, and user-friendly is important to achieving transparency in the use of digital technologies (Felzmann et al., 2020). Rubens (2024) mentioned that when transparency is missing, organisations may gain permission to use data, which would not have been granted if the owners had understood how such data would be used. This action in itself is unethical and should be avoided. Telkom (2023) noted that traceability, interactive, and comprehensible principles must be in place for transparency to be achieved. As such, it is essential to consider participants' backgrounds when communicating the intended purpose of data gathering to ensure proper comprehension.

Transparency in the ethical deployment of digital technologies in construction borders on open communication, clear documentation, and project participants' engagement. Using technologies like BIM and blockchain that promote collaboration among project participants can facilitate real-time discussion and feedback through open communication. Organisations can also employ open data policies that give project participants access to relevant data, aiding transparency in the decision-making process. Moreover, online surveys and related platforms can help project participants give feedback on the performance and concerns of adopted digital technologies. This can be very useful in monitoring technology progress and improvement where necessary. In line with monitoring technology performance, clear metrics must be implemented to measure impact, and regular reviews must be conducted with results shared with project participants. Based on this knowledge, transparency in the context of this study is conceptualised as 'having guidelines that ensure open communication, clear documentation, and project participants engagement in the use of digital technologies'. This ethical consideration is measured through:

1. Create guidelines to ensure the outcome of technology adoption is justifiable, interpretable, auditable, and visible.
2. Ensure that digital technology is adopted and the outcomes are traceable, interactive, and comprehensible.
3. Adopt digital technologies that promote open communication and data sharing.
4. Create guidelines for information sharing.
5. Promote detailed recording of the adoption process.
6. Ensure regulatory compliance in relation to information sharing.

Expected Success Outcomes of Ethical Use of Digital Technologies in Construction

Several benefits abound in the ethical use of digital technologies within construction organisations. Khodabakhshian (2024) has noted that as technology advances, the potential harms, biases, and discrimination associated with using these technologies in ethical and social contexts are becoming prominent. As such, being ethically aware when adopting these technologies can help avoid unethical conduct and provide construction projects that serve everyone within the society. Moreover, being ethically conscious when adopting emerging technologies can help build trust among project participants, leading to a long-lasting relationship (Joseph, 2023; Veinhardt, 2022). This can also prevent legal and financial risks associated with non-compliance with available regulations or wrongful use of individual data (Jemmy, 2024).

It has been observed that prioritising ethics in deploying digital technologies can give organisations a competitive advantage. Organisations can build a reputation for themselves and be known for their ethical considerations. This can prove attractive to clients and employees, creating a competitive advantage for themselves (Joseph, 2023). Strawser (2023) noted that in a crisis, an organisation's credibility can be its strongest asset; as such, clear and open communication and transparency are non-negotiables. This implies that construction organisations that prioritise ethical considerations in adopting digital technologies stand a better chance of overcoming crises on projects than those that do not. Moreover, long-term sustainability is possible as ethical considerations align with sustainable practices that consider efficient use of resources and elimination of waste in construction. Furthermore, by prioritising the social impacts of digital technologies, construction organisations can contribute to the well-being of communities and the environment (Martínez-Peláez et al., 2023).

Aside from the ethical considerations, employing digitalisation concepts in construction offers better project delivery and value for money (Aghimien et al., 2021). Health and safety can be achieved through robotics and automation of the construction process (Ruggiero et al., 2016). For instance, 3D printing of building components eliminates the need for onsite workers, reducing the risk of site accidents (Aghimien et al., 2020; Sakin & Kiroglu, 2017). Smart wearable devices with embedded sensors can also monitor workers' health issues and detect problems early, leading to effective health and safety on construction projects (Salento, 2017). Also, by adopting technologies, construction organisations can be more competitive (Kamaruddin et al., 2016).

Overall, prioritising ethical considerations in the deployment of digital technologies in construction organisations can lead to the following expected success outcomes.

1. Sustainable digitalisation of construction organisations.
2. Better competitiveness through enhanced reputation and trust.
3. Increased innovativeness.
4. Better risk mitigation.
5. Achieving long-term sustainability.
6. Better overall project performance.
7. Improved management of construction projects.
8. Becoming digital disruptor.

Theorising the Conceptualised Ethical Practice Framework for Sustainable Construction Digitalisation

The study theorises that for sustainable construction digitalisation, one that serves the present and future needs of construction organisations to be attained, four digitalisation dimensions must be considered, along with six critical ethical considerations. Attributes relating to an organisation's technology, people, process, and strategy in achieving digitalisation must be assessed along with accountability, equity and fairness, privacy, safety and security, sustainability, and transparency. By considering these ethical attributes in the quest for construction digitalisation, construction managers can effectively and ethically utilise their digital technologies and processes to develop new business models and more sustainable service delivery. Moreover, they can gain better competitiveness by attaining trust among their clients and employees while avoiding the risk of ethical breaches and efficiently managing construction projects. Figure 7.2 overviews the conceptualised ethical practice framework for sustainable construction digitalisation. Construction organisations that want to deploy emerging digital tools to attain sustainable construction digitalisation ethically, can adopt variables suitable to the conditions within their organisations. It is important to note that the proposed framework is designed to serve as a platform for further assessment within construction organisations.

Fig. 7.2 Conceptualised ethical practice framework for sustainable construction digitalisation. Source: Author's compilation (2024)

Summary

This chapter discussed the constituents of the conceptualised ethical practice framework for sustainable construction digitalisation. It explored the different dimensions of construction digitalisation (i.e. technology, people, process, and strategy), as noted in existing digitalisation frameworks and models. It also explored six ethical considerations that construction organisations must consider when employing these dimensions of construction digitalisation. This chapter also delineates some expected success outcomes of the ethical use of digital technologies in construction organisations. Finally, the chapter conceptualised a six-construct ethical practice model rooted in four digitalisation dimensions for the sustainable digitalisation of construction organisations. The next chapter concludes the book by summarising the key submissions and lessons.

References

Aghimien, D. O., Adegbemo, T. F., Aghimien, I. E., & Awodele, A. O. (2018). Challenges of sustainable construction: A study of educational buildings in Nigeria. *International Journal of Built Environment and Sustainability, 5*(1), 33–46.

Aghimien, D. O., Aigbavboa, C. O., Aghimien, L. M., Thwala, W. D., & Ndlovu, L. (2020). Making a case for 3D printing for housing delivery in South Africa. *International Journal of Housing Market and Analysis, 3*(4), 565–581.

Aghimien, D. O., Aigbavboa, C. O., Oke, A. E., & Thwala, W. D. (2021). *Construction digitalisation—A capability maturity model for construction organisations*. Routledge.

Aghimien, L. M., & Aigbavboa, C. (2024). A PLS-SEM analysis of emotional intelligence and construction organisation's performance nexus in South Africa. *Buildings, 14*, 3393. https://doi.org/10.3390/buildings14113393

Aguiar, T., Gomes, S. B., Cunha, P. R., & Silva, M. M. (2019). Digital transformation capability maturity model framework. In *IEEE 23rd international Enterprise distributed object computing conference (EDOC)*. IEEE. https://doi.org/10.1109/EDOC.2019.00016

Altimeter. (2024). *Eight factors of a successful digital transformation strategy*. Retrieved May 29, 2024, from https://prophet.com/2016/02/brief-the-opposite-approach-8-success-factors-of-digital-transformation/

Ashok, M., Madan, R., Joha, R., & Sivarajah, U. (2022). Ethical framework for artificial intelligence and digital technologies. *International Journal of Information Management, 62*, 1–17.

Bibby, L., & Dehe, B. (2018). Defining and assessing industry 4.0 maturity levels—Case of the defence sector. *Production Planning and Control, 29*(12), 1030–1043.

Bikitsha, L., Ngomane, N., & Mamafha, K. (2014). Understanding the use of emotional intelligence during the project leadership process: A case of PMs. *Journal of Leadership and Management Studies, 1*, 1.

Boström, E., & Celik, O. C. (2017). *Towards a maturity model for digital strategising—A qualitative study of how an organisation can analyse and assess their digital business strategy*. UMEA Universitet.

Braunack-Mayer, A. J., Street, J. M., & Palmer, N. (2012). Technology, ethics of: Overview. In *Encyclopedia of applied ethics* (pp. 321–327). Academic Press.

Dauchess, A. (2024). *Understanding the importance of ethics in information technology*. Retrieved December 1, 2024, from https://marymount.edu/blog/understanding-the-importance-

of-ethics-in-information-technology/#:~:text=The%20ethical%20issues%20in%20information,long%2Dterm%20benefits%20of%20technology
Day, G. S. (2011). Closing the marketing capabilities gap. *Journal of Marketing, 75*, 183–195.
De Carolis, A., Macchi, M., Negri, E., & Terzi, S. (2017). A maturity model for assessing the digital readiness of manufacturing companies. In H. Lödding et al. (Eds.), *APMS 2017. IFIP Advances in Information and Communication Technology*. Springer Verlag.
De Laat, P. B. (2021). Companies committed to responsible AI: From principles towards implementation and regulation? *Philosophy & Technology, 34*, 1135–1193.
Deloitte. (2018). *Digital maturity model - Achieving digital maturity to drive growth*. Retrieved May 29, 2024, from https://www2.deloitte.com/content/dam/Deloitte/global/Documents/Technology-Media-Telecommunications/deloitte-digital-maturity-model.pdf
Deloitte. (2022). *A framework for ethical tech*. Retrieved May 29, 2024, from https://action.deloitte.com/insight/3719/a-framework-for-ethical-tech
Dyk, L. V., & Schutte, C. S. L. (2012). Development of a maturity model for telemedicine. *South African Journal of Industrial Engineering, 23*(2), 61–72.
Eze, E. C., Aghimien, D. O., Aigbavboa, C. O., & Sofolahan, O. (2022). Building information modelling adoption for construction waste reduction in the construction industry of a developing country. *Engineering, Construction and Architectural Management, 31*(6), 2205–2223.
Felzmann, H., Fosch-Villaronga, E., Lutz, C., & Tamò-Larrieux, A. (2020). Towards transparency by design for artificial intelligence. *Science and Engineering Ethics, 26*(6), 3333–3361.
Gartner. (2016). *Digitalisation*. Retrieved December 11, 2023, from https://www.gartner.com/en/information-technology/glossary/digitalisation
Gerbert, P., Castagnino, S., Rothballer, C., Renz, A., & Filitz, R. (2016). *Digital engineering and construction: The transformative power of building information modelling*. The Bolton Consulting Group. Retrieved May 29, 2024, from https://www.bcg.com/publications/2016/engineered-products-infrastructure-digital-transformative-power-building-information-modeling
Gill, M., VanBoskirk, S., Evans, P. F., Nail, J., Causey, A., & Glazer, L. (2016). *The digital maturity model 4.0*. Benchmarks: Digital Business Transformation Playbook. Retrieved December 11, 2023, from www.forrester.com
Gökalp, E., & Martinez, V. (2021). Digital transformation capability maturity model enabling the assessment of industrial manufacturers. *Computers in Industry, 132*, 1–23.
Hussain, S., Mahmood, K., Khan, M. K., Chen, C., Alzahrani, B. A., & Chaudhry, S. A. (2022). Designing secure and lightweight user access to drones for smart city surveillance. *Computer Standards & Interfaces, 80*, 1–11.
Institute of Chartered Accountants in England and Wales. (2019). *New technologies, ethics and accountability*. Retrieved May 29, 2024, from https://www.icaew.com/-/media/corporate/files/technical/technology/thought-leadership/new-technologies-ethics-and-accountability.ashx
International Energy Agency. (2019). Global status report for buildings and construction 2019 : International Energy Agency. Retrieved August 8, 2023, from https://shorturl.at/dmzGZ
Jemmy, S. (2024). Integration of business ethics in managerial decisions implications for companies' long-term growth. *International Journal of Social and Education, 1*(6), 1689–1702.
Joseph, A. (2023). The importance of business ethics: Building trust and sustainability. *Business Studies Journal, 15*(S2), 1–2.
Kamaruddin, S., Mohammed, M. F., & Mahbub, E. (2016). Barriers and impact of mechanisation and automation in construction to achieve better quality products, Malaysia. *Procedia - Social and Behavioural Sciences, 222*, 111–120.
Khodabakhshian, A. (2024). Ethics-aware application of digital technologies in the construction industry. In S. Chiodo, D. Kaiser, J. Shah, & P. Volonté (Eds.), *Improving technology through ethics. Springer Briefs in Applied Sciences and Technology*. Springer.
Lemperos, X. (2024). *How does BIM and LCA fit together?* Retrieved August 12, 2024, from https://anavitor.ai/en/ecobuild-lca-perspectives-for-the-built-environment/how-does-bim-and-lca-fit-together#:~:text=Anavitor%20LCA%20%2D%20a%20two%2Dway%20tool%20

for%20BIM%20and%20LCA&text=Anavitor%20LCA%20is%20a%20promising,for%20architects%20working%20with%20LCA

Leyh, C., Köppel, K., Neuschl, S., & Pentrack, M. (2021). Critical success factors for digitalisation projects. In *Proceedings of the 16th conference on computer science and intelligence systems* (pp. 427–436). IEEE. https://doi.org/10.15439/2021F122

Marrero, M., Wojtasiewicz, M., Martínez-Rocamora, A., Solís-Guzmán, J., & Alba-Rodríguez, M. D. (2020). BIM-LCA integration for the environmental impact assessment of the urbanisation process. *Sustainability, 12*(10), 4196.

Martínez-Peláez, R., Ochoa-Brust, A., Rivera, S., Félix, V. G., Ostos, R., Brito, H., Félix, R. A., & Mena, L. J. (2023). Role of digital transformation for achieving sustainability: Mediated role of stakeholders, key capabilities, and technology. *Sustainability, 15*(14), 11221.

McLaughlin, S. A. (2017). Dynamic capabilities: Taking an emerging technology perspective. *International Journal of Manufacturing Technology and Management, 31*(1-3), 62–81.

Milesight. (2024). *Smart building sensors: A comprehensive guide for facility managers*. Retrieved August 12, 2024, from https://www.milesight.com/company/blog/iot-smart-building-sensors#:~:text=Benefits%20of%20Installing%20Smart%20Building%20Sensors,-Energy%20Efficiency&text=Sensors%20optimize%20the%20use%20of,on%20investment%20for%20sensor%20installation

Morin, X., & Romero-Torres, A. (2024). How does building information modelling influence decision-making process in the project design? An input, process and output analysis. *Project Leadership and Society, 5*, 1–12.

Newman, M. (2017). *Digital maturity model (DMM): A blueprint for digital transformation*. TM Forum White Paper. Retrieved December 11, 2024, from www.tmforum.org

Oke, A. E., Aigbavboa, C., Aghimien, D. O., & Koloko, N. (2018). Challenges of digital collaboration in the south African construction industry. In *Proceedings of the international conference on industrial engineering and operations management* (pp. 2472–2482).

Olorunfemi, O. L., Amoo, O. O., Atadoga, A., Fayayola, O. A., Abrahams, T. O., & Shoetan, P. O. (2024). Towards a conceptual framework for ethical AI development in IT systems. *Computer Science & IT Research Journal, 5*(3), 616–627.

Palm, E., & Hansson, S. O. (2006). The case for ethical technology assessment (eTA). *Technological Forecasting and Social Change, 73*, 543–558.

Prathomwong, P., & Singsuriya, P. (2022). Ethical framework of digital technology, artificial intelligence, and health equity. *Asia Social Issues, 15*(5), 1–14.

Project Management Body of Knowledge (PMBOK). (2017). *A guide to project management body of knowledge* (6th ed.). Project Management Institute.

Rubens, P. (2024). *What is digital ethics?* Retrieved August 12, 2024, from https://charitydigital.org.uk/topics/what-is-digital-ethics-7593

Ruggiero, A., Salvo, S., & Laurent, C. (2016). *Robotics in construction*. Worcester Polytechnic Institute.

Sahu, M. (2023). *5+ advantages of blockchain technology with real-life examples*. Retrieved August 12, 2024, from https://www.upgrad.com/blog/advantages-of-blockchain-technology/

Sakin, M., & Kiroglu, Y. C. (2017). 3D printing of buildings: Construction of the sustainable houses of the future by BIM. *Energy Procedia, 134*, 702–711.

Salento, A. (2017). Digitalisation and the regulation of work: Theoretical issues and normative challenges. *Artificial Intelligence and Society, 33*, 369–378.

Sawhney, A., & Knight, A. (2023). *Digitalisation in construction report 2023*. Retrieved August 12, 2024, from https://www.rics.org/content/dam/ricsglobal/documents/research/Digitalisation%20in%20construction%202023_final.pdf

Schoentgen, A., & Wilkinson, L. (2021). Ethical issues in digital technologies. In *23rd biennial conference of the international telecommunications society (ITS): Digital societies and industrial transformations: Policies, markets, and technologies in a post-Covid world*. International Telecommunications Society (ITS).

Sheikhshoaei, F., Naghshineh, N., Alidousti, S., & Nakhoda, M. (2018). Design of a digital library maturity model (DLMM). *The Electronic Library, 36*(4), 607–619.

Strawser, B. (2023). *Crisis management: Strengthening stakeholder trust.* Retrieved August 12, 2024, from https://bryghtpath.com/crisis-management-stakeholder-trust/

Struk, V. (2024). *How regular IoT firmware updates can prevent security breaches.* Retrieved August 12, 2024, from https://relevant.software/blog/iot-firmware-update/#:~:text=Firmware%20updates%20often%20include%20patches,on%20IoT%20device%20security%20concepts

Teece, D. J. (2010). Business models, business strategy and innovation. *Long Range Planning, 43*, 172–194.

Telkom. (2023). *Digital ethics framework.* Retrieved May 29, 2024, from https://group.telkom.co.za/documents/sustainability/Digital_Ethics_Framework.pdf

Valdez-de-Leon, O. (2016). A digital maturity model for telecommunications service providers. *Technology Innovation Management Review, 6*(8), 19–32.

Varona, D., & Suárez, J. L. (2022). Discrimination, bias, fairness, and trustworthy AI. *Applied Sciences, 12*(12), 5826.

Veinhardt, J. (2022). The dilemma of postmodern business ethics: Employee reification in a perspective of preserving human dignity. *Frontiers in Psychology, 373*, 813255.

World Green Building Council. (2019). *Bringing embodied carbon upfront.* Retrieved August 12, 2024, from https://worldgbc.org/advancing-net-zero/embodied-carbon/

Zostant, M., & Chataut, R. (2023). Privacy in computer ethics: Navigating the digital age. *Computer Science and Information Technologies, 4*(2), 183–190.

Chapter 8
Conclusion

Introduction

The general objective of this book was to provide a conceptualised ethical practice framework that will engender sustainable digitalisation within construction organisations. This was deemed necessary due to the significant impact emerging digital technologies are making in the construction industry and the ethical concerns integrating these technologies are raising. To achieve this overarching objective of the book, diverse key concepts like digitalisation, ethics, and sustainability were explored. The anatomy of the construction industry was X-ray vis-à-vis the industry's readiness to be digitalised. Also, the sustainable construction and digitalisation nexus and the relationship between ethics and digitalisation were explored. Furthermore, the study draws from existing digital ethics and ethics of technology frameworks to develop an ethical practice framework for sustainable construction digitalisation. Consequently, this last chapter concludes the book by summarising the key areas covered and signposting directions for the ethical use of emerging digital technologies in the quest for sustainable construction digitalisation.

Conclusion

The fourth industrial revolution, characterised by physical, biological, and digital technologies, has continued to revolutionise the way different sectors operate, including construction. The constant development of technologies coupled with the benefits these technologies propose in addressing age-long construction issues has led to a continuous clamour for integrating emerging technologies in the construction industry. This is because, globally, the construction industry has been berated for its slow adoption of emerging technologies, leading to poor innovativeness in

O. Oladinrin et al., *Ethical Practices for Sustainable Construction Digitalisation*, https://doi.org/10.1007/978-3-032-00151-1_8

developed and developing countries. Although the construction industry globally plays a crucial role in socio-economic development and infrastructure delivery, the industry in many countries is characterised by poor project delivery due to cost, time, quality, sustainability, client and user satisfaction, health and safety, among others. These problems are the evidence of multifaceted problems facing the construction industry in different countries. These include cash flow and financial issues, poor management, poor sustainability thinking, issues with supply chains, poor image of the industry, the industry's resistance to change and slow adoption of technologies, among others. Emerging digital technologies offer solutions to address these problems.

Building Information Modelling (BIM), which has continued to gain prominence in the industry, offers faster project delivery through efficient collaboration. It also allows effective cost delivery of projects through efficient planning, elimination of clashes and errors in design, and proper management to eliminate waste and promote sustainable construction (SC). When used with cloud-based technologies and the Internet of Things (IoT), it offers remote access capabilities and real-time collaboration, which further improves project efficiency. Cloud computing further offers effective gathering and storage of construction data. At the same time, big data analytics (BDA) can critically analyse big data from construction projects to make informed decisions and plan effectively. Automation and robotics such as three-dimensional (3D) printing, autonomous robots, and drones offer faster, safer, and more SC through machines. By deploying autonomous robots and drones, safety on construction sites can also be achieved as workers are removed from harm's way. This further promotes the social sustainability of construction projects.

Smart wearables like smart wristwatches, helmets, boots, vests, etc. are revolutionising how workers' health and safety are being monitored on construction sites. These technologies are helping to improve workers' health, safety, and productivity while reducing absenteeism. Again, 3D printing has been noted for its sustainability characteristics, as it can use sustainable materials in printing while reducing waste, which has been a major sustainability issue for the global construction industry. Blockchain technology offers transparency, traceability, and accountability in the supply chain. It also offers opportunities to confirm the sustainability of materials that have been marked as sustainable by the supplier. This allows for effective supply chain management in construction.

Smart technologies in buildings like LED lighting, smart water metres, and HVAC, among others, offer energy efficiency and CO_2 reduction in construction. Moreover, digital technologies like AI, IoT, blockchain, and others offer sustainability benefits like better monitoring and control of resources, easy selection of suitable green material, better decision-making, efficient supply chain, better communication and collaboration, and waste elimination on construction projects. Evidently, utilising these technologies effectively can lead to the digitalisation of construction organisations. These technologies offer efficiency in organisations' service delivery and the development of new business models that could lead to new revenue generation.

Despite the existence of these emerging digital technologies, they come with their ethical challenges. Generally, the construction industry has been known for unethical practices that have necessitated diverse professional, ethical guidelines designed to ensure morals and ethical behaviours among professionals. With the call for the deployment of digital technologies in the construction industry, ethical use has been a critical area needing attention. Ethics of technology, which entails the moral principles and standards that guide the ethical conduct and responsibilities of individuals, organisations, and governments in developing and using emerging technologies, has become a crucial aspect worth exploring if the digitalisation being sought after is to be sustained. Moreover, integrating emerging digital technologies involves capturing and managing massive volumes of data, generating questions about how this information is stored, shared, and used. Ethical concerns about data ownership and potential misuse or breaches necessitate strong frameworks and policies to protect stakeholder interests. Other ethical concerns regarding emerging technologies have revolved around poor accountability and responsibility, privacy and security issues, lack of transparency, regulatory compliance problems, accuracy problems, and ethical washing, among others. Addressing these issues in the construction industry becomes timely as no evidence of such a study proposing a pathway for achieving ethical use of digital technologies with construction organisations.

Drawing from existing ethics of technology and digital ethics frameworks, various ethical considerations were discovered to be crucial to deploying digital technologies in construction organisations. The most recurring ethical considerations were identified: accountability, equity and fairness, privacy, safety and security, sustainability, and transparency. For construction organisations to effectively digitalise their services and processes through emerging digital technologies and sustain this digitalisation, accountability in terms of having clearly defined guidelines that detail the responsibilities and liabilities of various parties in deploying emerging technologies is necessary. Aside from defining responsibilities and liabilities, construction organisations must put measures in place to ensure that digital technologies employed are inclusive and fair to all participants affected by the delivered construction project. They must also ensure that the decisions made based on the results of these digital technologies do not exhibit bias against any demographic group.

Furthermore, privacy is essential as this has been a serious ethical challenge for successfully deploying technologies like AI. Construction organisations must put clear guidelines in place to ensure ethical gathering, storing, analysing, and destroying data gathered through digital technologies. Moreover, they must ensure project participants have autonomy over their personal information, and how they want it to be used. This also applies to the safety and security of individuals and digital assets from unforeseen activities such as unauthorised access, data breaches, and cyber threats that might cause harm. Putting measures in place to ensure safety is important for organisations seeking sustainable digitalisation. These organisations must also be conscious of some sustainability criteria that can help enforce their resolve for ethical use of digital technologies. They need to have guidelines that consider the environmental, social, and economic impact of digital technology deployments while ensuring that these technologies are used responsibly without creating issues for future generations. Finally, transparency is essential for the ethical deployment of digital technologies in construction organisations. Therefore, these organisations must have guidelines that ensure open communication, clear documentation, and project participants' engagement in the use of digital technologies.

In the quest for sustainable digitalisation, the existing review revealed that construction organisations must consider the dimensions of technology, people, process, and strategy. While the technology aspect encompasses the available and applicable digital technologies along with other associated implementation factors, the people dimension relates to all activities that directly relate to people, either employees, employers, or customers. The process dimension includes organisational and managerial processes that coordinate activities to deliver better construction products and services. Lastly, the strategy is seen as measures put in place to sense opportunities and threats in emerging digital technologies, seize them, and use them to transform organisational activities. Understanding these four critical dimensions can lead to the digitalisation of construction organisations; to sustain this digitalisation and make it effective, it is necessary to prioritise the aforementioned ethical considerations. As such, this book proposes a four-construct digitalisation and six ethical consideration framework that is designed to offer construction organisations sustainable construction digitalisation, better competitiveness through enhanced reputation and trust, increased innovativeness, better risk mitigation, achieving long-term sustainability, better overall project performance, improved management of construction projects, and the opportunity to become a digital disruptor.

Based on the above, the study achieved its set objective of delivering a conceptualised framework of ethical practices for sustainable construction digitalisation. However, it is imperative to note that the proposed ethical practice framework developed and presented in this book is designed to serve as a platform for further assessment within construction organisations. In adopting the proposed framework, it is important to test its applicability within the local context and adjust to suitability and applicability as appropriate.

Contribution to Knowledge

The contribution of this book to the existing discourse on construction digitalisation can be seen in two different folds—practical and theoretical.

Practical Contribution

Practically, the book showcases guidelines for construction organisations, policymakers, and industry practitioners to attain construction digitalisation that can sustained through reasonable ethical considerations. For the management of construction organisations, the study proposed six ethical considerations essential for the effective diffusion of digital technologies into construction activities. Since the construction industry is known as a consumer of technologies, ethically using these technologies in the delivery of their projects is important. As such, by promoting accountability, they can ensure that responsibilities are shared accordingly and that the right individuals are responsible for the different technologies employed and their outcomes. Moreover, equality and fairness will help ensure that all participants can use the technologies being introduced and that the decisions made from using these technologies are fair and just. Organisations must also ensure privacy by employing measures put in place to safeguard crucial project information and personal data. This is essential as the issue of privacy, security, and safety has been a topic of debate since the advent of digital technologies. Initiatives such as encryption, authentication, and access controls can help promote security on construction projects while ensuring adherence to existing regulations that protect individuals' rights to control their personal information, which is crucial for privacy. Moreover, advancements in emerging technologies like cloud computing offer opportunities for additional layers of security for organisation and project data. Effectively using such technology while sticking to existing best practices and legal frameworks can further improve information safety and security, leading to a better competitive advantage for the organisation. No doubt sustainability is essential for organisations seeking to sustain their attained digitalisation. Adopting emerging digital tools that promote sustainability in construction projects is essential. These technologies can help promote the environmental, social, and economic impact of construction projects. Also, it is essential for organisations to assess how their adoption of these technologies' impacts on the immediate environment, individuals, and societies. This will help them identify their sustainability impact and create measures to promote construction projects that meet the needs of the present as well as future generations. Also, it is important to be open in terms of communication about how technology will be used and how decisions will be made based on the outcomes of using these technologies. Being transparent is crucial to having project participants buy into the adoption of emerging technologies.

The proposed framework also offers guidelines for policymakers within the construction industry. Through this framework, construction industry regulatory and professional bodies and other relevant policymakers can understand the core ethical considerations that their members and industry practitioners must abide by. Measures can then be put in place to enforce these ethical considerations and align existing policies to meet these ethical considerations. Moreover, industry practitioners, including stakeholders, can benefit from the proposed framework, especially in their understanding of the ethical use of digital technologies and how it can impact their activities. Aside from organisations and industry regulatory bodies enforcing the ethical use of digital technologies, professionals also play a critical role in upholding and sticking to the ethical practices in using these technologies. As such, the proposed framework will guide these individuals in their use of these technologies.

Theoretical Contribution

The importance of ethical use of digital technologies has continued to be a topical issue among academics, practitioners, and stakeholders in diverse sectors. The continuous improvement in AI and related technologies has created some apprehension about how these technologies are used and how they impact individuals, society, and the environment. As such, studies have explored the ethics of technology and digital ethics within diverse sectors. However, the review of these studies revealed that such studies are deficient in the construction domain despite the unethical nature of the industry, as noted in existing studies. Therefore, this book is timely as it provides a collection of knowledge for academics, organisations, and regulatory bodies to use in their quest to achieve sustained construction digitalisation through proper ethical practices. The book also serves as an excellent platform for built environment researchers in academics and practice seeking to explore the ethical deployment of emerging digital technologies in construction. It gives a basis for further research to be conducted in this area, particularly in countries where such study is non-existent. Based on the conceptualised ethical practice framework for sustainable construction digitalisation proposed in Chap. 7, this book presents the theory that to achieve sustainable construction digitalisation, better competitiveness, and overall improved project performance, among other benefits, construction organisations, and practitioners must prioritise the technology, people, process, and strategy dimensions of digitalisation. These must be shaped by six core ethical considerations: accountability, equity and fairness, privacy, safety and security, sustainability, and transparency.

Recommendations

The review of existing works in the book revealed that despite being a crucial part of effective technology deployment, ethics in using digital technologies have not been a topical issue in the construction domain. While existing studies with engineering and related fields show a growing rise in technology usage, emphasis on the ethical use of technology within construction is still very scant. As such, it is essential for the management of construction organisations and project participants to prioritise ethical practices in their selection of emerging technologies and in the usage of these technologies. Moreover, by prioritising the six ethical considerations proposed in this book, these organisations can build sustainable business practices and trust in their projects and clients, which can transcend into a competitive advantage. Also, through ethical practices, they can ensure compliance with existing best practices, regulations, and legislation, particularly in privacy and the safety of individuals' information. Construction organisations and industry participants must understand that by employing ethical practices, they can mitigate potential harm and avoid huge costs associated with harm caused by the selected digital technology. Moreover, they can show consistency within their organisation by adopting proper ethical codes for the use of emerging technologies.

Policymakers and regulatory bodies are also challenged to review existing regulations and ensure that organisations and professionals give adequate consideration to the ethical use of technologies. The proposed six ethical considerations can serve as a viable foundation in reviewing existing regulations around the use of digital technologies in the delivery of construction projects. Also, it is essential to put measures in place to enforce these regulations and ensure they are adhered to. In the context of accountability, construction organisations and professionals must be accountable for the decisions made using emerging technologies. Where negative outcomes are evident, measures for redress must be clear in the regulations to ensure organisations take responsibility. This should also include a clear indication of the need for competency in deploying these technologies. Construction organisations must be bound by regulations that clearly indicate the need for individuals with the right competence to operate and interpret these digital technologies to make informed decisions.

The conceptualised ethical practice framework for sustainable construction digitalisation presented in this book provides a broad view of the dimensions and ethical practices of digitalisation as well as the measurement variables needed. For effective use, organisations, practitioners, and researchers must test the suitability and applicability of the conceptualised framework further in the context of the country where it is to be applied.

MIX
Papier aus verantwortungsvollen Quellen
Paper from responsible sources
FSC® C105338
FSC
www.fsc.org

If you have any concerns about our products,
you can contact us on
ProductSafety@springernature.com

In case Publisher is established outside the EU,
the EU authorized representative is:
Springer Nature Customer Service Center GmbH
Europaplatz 3, 69115 Heidelberg, Germany

Printed by Libri Plureos GmbH
in Hamburg, Germany